George Nichols Packer

Our Calendar

The Julian calendar and its errors. How corrected by the Gregorian.

George Nichols Packer

Our Calendar
The Julian calendar and its errors. How corrected by the Gregorian.

ISBN/EAN: 9783337018788

Printed in Europe, USA, Canada, Australia, Japan

Cover: Foto ©berggeist007 / pixelio.de

More available books at **www.hansebooks.com**

OUR CALENDAR.

The Julian Calendar and Its Errors.

HOW CORRECTED BY THE GREGORIAN.

Rules For Finding the Dominical Letter,

AND THE DAY OF THE WEEK OF ANY EVENT FROM THE
DAYS OF JULIUS CÆSAR 46 B. C. TO THE YEAR OF
OUR LORD FOUR THOUSAND—A NEW AND EASY
METHOD OF FIXING THE DATE OF EASTER.

HEBREW CALENDAR;

SHOWING THE CORRESPONDENCE IN THE DATE OF
EVENTS RECORDED IN THE BIBLE WITH OUR
PRESENT GREGORIAN CALENDAR.

ILLUSTRATED BY VALUABLE TABLES AND CHARTS.

BY REV. GEORGE NICHOLS PACKER,
CORNING, N. Y.

TO

HON. HENRY W. WILLIAMS,

JUSTICE OF THE SUPREME COURT OF PENNSYLVANIA

WHOM

I HAVE FOUND A TRUE FRIEND IN POVERTY AND IN SICKNESS,

AND

FROM WHOM I HAVE RECEIVED WORDS OF ENCOURAGEMENT

AND COMFORT DURING MANY YEARS OF ADVERSITY,

AND AT

WHOSE SUGGESTION THIS LITTLE VOLUME HAS BEEN WRITTEN,

AND BY

WHOSE ASSISTANCE IT IS NOW PUBLISHED,

THIS

HUMBLE VOLUME IS DEDICATED

AS A

TRIBUTE OF RESPECT

BY THE

AUTHOR.

PREFACE.

MANY years ago, while engaged in teaching, the writer of this little volume was in the habit of bringing to the attention of his pupils a few simple rules for finding the dominical letter and the day of the week of any given event within the past and the present centuries ; further than this he gave the subject no special attention.

A few years ago, having occasion to learn the day of the week of certain events that were transpiring at regular intervals on the same day of the same month, but in different years, he was led to investigate the subject more thoroughly, so that he is now able to give rules for finding the dominical letter and the day of the week of any event that has transpired or will transpire, from the commencement of the Christian era to the year of our Lord 4,000, and to explain the principles on which these rules rest. When the investigations were entered upon he had no thought of writing a book ; but having been laid aside from active labor by ill health, he found relief from the despondency in which sickness and poverty plunged him by pursuing the study of the calendar, its history, and the method of disposing of the fraction of a day found in the time required for the revolution of the Earth in its orbit about the Sun.

He became so much interested in the study of this subject that he frequently spoke of it to friends and acquaintances whom he met. On one occasion, while speaking to Hon. H. W. Williams about some of the curious results of the process by which the coincidence of the solar and the civil year is preserved, it was suggested to him that he should put the story of the calendar, its correction by Gregory, and the theory and results of intercalation, in writing. It was urged that this would give increased interest to the study, help the writer to forget his pains, and probably enable him to realize a little money from the sale of his work to meet pressing wants. Acting upon this suggestion, an effort has been made to put into this little volume some of the most interesting facts relating to the origin, condition, and practical operation of the calendar now in use; together with rules for finding the day of the week on which any given day of any month has fallen or will fall during four thousand years from the beginning of our era.

The writer does not claim absolute originality for all that appears in the following pages; on the contrary, he has made free use of all the materials that came within his reach relating to the history of the calendar and the work of its correction by Gregory. These ma-

terials, together with his own calculations, he has arranged in accordance with a plan of his own devising,
so that the outline and the execution of the work may
be truly said to be original. Of its value the world
must judge. It has been prepared in weakness of body
and in suffering, which have been to some extent re-
lieved by the mental occupation thus afforded, but
which may have nevertheless left their impress on the
work. But let it be read before pronouncing judgment
upon it. Cicero could infer the littleness of the Hebrew
God from the smallness of the territory he had given
his people. To whom Kitto replies: "The interest and
importance of a country arise, not from its territorial
extent, but from the men who form its living soul;
from its institutions bearing the impress of mind and
spirit, and from the events which grow out of the character and condition of its inhabitants." So the value
of a book does not consist in the size and number of
its pages, but from the knowledge that may be gained
by its perusal. THE AUTHOR.

PREFACE

TO THE REVISED EDITION.

———

Soon after the publication of the former edition of this work, it was suggested that a chapter be added on Easter; rules for fixing its date, aad also church festivals that depended upon the date of Easter. It was suggested that this would add very much to the value of the work, if so presented as to be brought within the comprehension of ordinary minds. Knowing that the determination of Easter was an affair of considerable nicety and complication, and had had the attention of our best minds, and they had failed so to present it, that even among scholarly men, probably not one in a hundred was able to determine its date without referring to tables prepared for that purpose the author of this work felt as though he was hardly competent for the task. Nevertheless it was undertaken, and the work has been revised and enlarged by a Chapter on the Peculiarities of the Roman Calendar, another on fixing the date of events prior to the Christian era, and a third part on Easter, church festivals, and the Hebrew Calendar. In the opinion of the author, the rules for determining the date of Easter are so simplified by his new method that any person of ordinary intelligence may understand them. How well he has succeeded the public will decide.　　　　G. N. P.

CONTENTS.

— ◆ —

PART FIRST.

DEFINITIONS—HISTORY.

PART SECOND.

MATHEMATICAL.

PART THIRD.

CYCLES—JULIAN PERIOD—EASTER.

OUR CALENDAR.

PART FIRST.

DEFINITIONS. HISTORY.

CHAPTER I.

DEFINITIONS.

a—A Calendar is a method of distributing time into certain periods adapted to the purposes of civil life, as hours, days, weeks, months, years, etc.

b—The only natural divisions of time are the solar day, the solar year, and the lunar month.

c—An hour is one of the subdivisions of the day into twenty-four equal parts.

d—The true solar day is the interval of time which elapses between two consecutive returns of the same terrestrial meridian to the Sun, the mean length of which is twenty-four hours.

e—The week is a period of seven days, having no reference whatever to the celestial motions, a circumstance to which it owes its unalterable uniformity.

f—The lunar month is the time which elapses between two consecutive new or full moons, and was used in the Roman calendar until the time of Julius Cæsar, and consists of 29d, 12h, 44m, 2.87s.

g—The calendar month is usually employed to denote an arbitrary number of days approaching a twelfth part of a year, and has now its place in the calendar of nearly all nations.

h—The year is either astronomical or civil. The solar astronomical year is the period of time in which the Earth performs a revolution in its orbit about the sun or passes from any point of the ecliptic to the same point again, and consists of 365 days, 5 hours, 48 minutes and 49.62 seconds of mean solar time. Appendix A.

i—The civil year is that which is employed in chronology, and varies among different nations, both in respect of the seasons at which it commences and of its subdivisions.

CHAPTER II.

HISTORY OF THE DIVISIONS OF TIME AND THE OLD ROMAN CALENDAR.

Day—The subdivision of the day into twenty-four parts or hours has prevailed since the remotest ages, though different nations have not agreed either with respect to the epoch of its commencement or the manner of distributing the hours. Europeans in general, like the ancient Egyptians, place the commencement of the civil day at midnight; and reckon twelve morning hours from midnight to midday and twelve evening hours from midday to midnight. Astronomers, after the example of Ptolemy, regarded the day as commencing with the Sun's culmination, or noon, and find it most convenient for the purpose of computation to reckon through the whole twenty-four hours. Hipparchus reckoned the twenty-four hours from midnight to midnight.

The Roman day, from sunrise to sunset, and the night, from sunset to sunrise, were each divided at all seasons of the year into twelve hours, the hour being uniformly one-twelfth of the day or the night, of course, varied in length with the length of the day or night at different seasons of the year.

Week—Although the week did not enter into the calendar of the Greeks, and was not introduced at Rome till after the reign of Theodosius, A. D. 292, it has been employed from time immemorial in almost all Eastern countries ; and as it forms neither an aliquot part of a year nor of the lunar months, those who reject the Mosaic recital will be at a loss to assign to it an origin having much semblance of probability. In the Egyptian astronomy the order of the planets, beginning with the most remote, is Saturn, Jupiter, Mars, the Sun, Venus, Mercury, the Moon. Now, the day being divided into twenty-four hours, each hour was consecrated to a particular planet, namely : One to Saturn, the following to Jupiter, third to Mars, and so on according to the above order ; and the day received the name of the planet which presided over its first hour. If, then, the first hour of a day was consecrated to Saturn, that planet would also have the 8th, the 15th and the 22d hours ; the 23d would fall to Jupiter, the 24th to Mars, and the 25th or the first hour of the second day would belong to the Sun. In like manner the first hour of the third day would fall to the Moon, the first hour of the fourth to Mars, of the fifth to Mercury, of the sixth to Jupiter and the seventh to Venus. The cycle being completed, the first hour of the eighth day would again return to Saturn and all the others succeed in the same order. See table on the 17th page.

It will be seen by the table, and it is also recorded by Dio Cassius, of the second Century, that the Egyptian week commenced with Saturday. On their flight from Egypt the Jews, from hatred to their ancient oppressors, made Saturday the last day of the week. It is stated that the ancient Saxons borrowed the week from some Eastern nation, and substituted the names of their own divinities for those of the gods of Greece. The names of the days are here given in Latin, Saxon and English. It will be seen that the English names of the days are derived from the Saxon.

LATIN.	SAXON.	ENGLISH.
Dies Solis.	Sun's Day.	SUNDAY.
Dies Lunae.	Moon's Day.	MONDAY.
Dies Martis.	Tiw's Day.	TUESDAY.
Dies Mercurii.	Woden's Day.	WEDNESDAY.
Dies Jovis.	Thor's Day.	THURSDAY.
Dies Veneris.	Friga's Day.	FRIDAY.
Dies Saturni.	Seterne's Day.	SATURDAY.

Month—The ancient Roman year contained but ten months and is indicated by the names of the last four. September from Septem, seven; October from Octo, eight; November from Novem, nine, and December from Decem, ten; July and August were also denominated Quintilis and Sextilis, from Quintus five, and Sex, six.

Quintilis was changed to July in honor of Julius Cæsar, who was born on the 12th of that month 98 B. C. Sextilis was changed to August by the Roman Senate to flatter Augustus on his victories about 8 B. C. In the reign of Numa Pompilius, about 700 B. C., two months were added to the year, January at the beginning, and February at the end of the year. This arrangement continued till 450 B. C., when the Decemvirs (ten magistrates) changed the order, placing February after January, making March the third instead of the first month of the Roman year.

Year—If the civil year correspond with the solar the seasons of the year will always come at the same period. But if the civil year is supposed to be too long (as is the case in the Julian year) the seasons will go back proportionately ; but if too short they will advance in the same proportion. Now, as the ancient Egyptians reckoned thirty days to the month invariably, and to complete the year, added five days, called supplementary days, their year consisted of 365 days.

They made use of no intercalation, and by losing one-fourth of a day every year, the commencement of the year went back one day in every period of four years, and consequently made a revolution of the seasons in 1460 years. Hence the Egyptian year was called a vague or erratic year because the first day of the year in the course of 1460 years wandered, as it were, over all the seasons. Therefore 1460 Julian years of 365¼ days each are equal to 1461 Egyptian years of 365 days each.

The ancient Roman year consisted of twelve lunar months, of twenty-nine and thirty days alternately,

which equals 354 days; but a day was added to make
the number odd, which was considered more fortunate,
so that the year consisted of 355 days.

This differed from the solar year by ten whole days
and a fraction; but to restore the coincidence, Numa
ordered an additional or intercalary month to be inserted every second year between the 23d and 24th of
February, consisting of twenty-two and twenty-three
days alternately, so that four years contained 1465 days,
and the mean length of the year was consequently 366¼
days, so that the year was then too long by one
day.

As the error amounted to twenty-four days in as
many years, it was ordered that every third period of
eight years, instead of containing four intercalary
months, two of twenty-two and two of twenty-three
days, amounting in all to ninety days, should contain
only three of those months of twenty-two days each,
amounting to sixty-six days, thereby suppressing twenty-four days in as many years, reducing the mean length
of the year to 365¼ days.

Had the intercalations been regularly made the concurrence of the solar and the civil year would have
been preserved very nearly. But its regulation was left
to the pontiffs, who, to prolong the term of a magistra-
cy or hasten an annual election, would give to the intercalary month a greater or less number of days, and
consequently the calendar was thrown into confusion,
so that in the time of Julius Cæsar there was a discrepancy between the solar and the civil year of about three
months; the winter months being carried back into
autumn and the autumnal into summer.

A table of the.order and the names of the planets in the Egyptian astronomy illustrating the origin of the names of the days of the week :

Saturn, Saturday.	Jupiter, Thursday.	Mars, Tuesday.	Sun, Sunday.	Venus, Friday.	Mercury, Wednesday.	Moon, Monday.
1	2	3	4	5	6	7
8	9	10	11	12	13	14
15	16	17	18	·19	20	21
22	23	24	1	2	3	4
5	6	7	8	9	10	11
12	13	14	15	16	17	18
19	20	21	22	23	24	1
2	3	4	5	6	7	8
9	10	11	12	13	14	15
16	17	18	19	20	21	22
23	24	1	2	3	4	5
6	7	8	9	10	11	12
13	14	15	16	17	18	19
20	21	22	23	24	1	2
3	4	5	6	7	8	9
10	11	12	13	14	15	16
17	18	19	20	21	22	23
24	1	2	3	4	5	6
7	8	9	10	11	12	13
14	15	16	17	18	19	20
21	22	23	24	1	2	3
4	5	6	7	8	9	10
11	12	13	14	15	16	17
18	19	20	21	22	23	24

CHAPTER III. .

HISTORY OF THE REFORMATION OF THE CALENDAR BY JULIUS CÆSAR.

In order to put an end to the disorders arising from the negligence or ignorance of the pontiffs, Julius Cæsar, 46 B. C., abolished the use of the lunar year and the intercalary month, and regulated the civil year entirely by the Sun. With the advice and assistance of the astronomers, especially Sosigenes of Alexandria, he fixed the mean length of the year at 365¼ days, and decided that there should be three consecutive years of 365 days, and a fourth of 366.

In order to restore the vernal equinox to the 24th of March, the place it occupied in the time of Numa, two months, together consisting of 67 days, were inserted between the last day of November and the first day of December of that year. An intercalary month of 23 days had already been added to February of the same year according to the old method, so that the first Julian year commenced with the first day of January, 45 years before Christ, and 709 from the foundation of Rome, making the year A. U. C. 708 to consist of the prodigious number of 445 days, (i. e. 355+23+67= 445). Hence it was called by some the year of confusion ; Macrobius said it should be named the last year of confusion.

There was also adopted at the same time a more commodious arrangement in the distribution of the days throughout the several months. It was decided to give to January, March, May, July, September and November each thirty-one days ; and the other months

thirty, excepting February, which in common years
should have but twenty-nine days, but every fourth
year thirty; so that the average length of the Julian
year was 365¼ days.

Augustus Cæsar interrupted this order by taking
one day from February, reducing it to twenty-eight
and giving it to August, that the month bearing his
name should have as many days as July, which was
named in honor of his .great-uncle, Julius. In order
that three months of thirty-one days might not come
together, September and November were reduced to
thirty days, and thirty-one given to October and December.

In the Julian calendar a day was added to February
every fourth year, it being the shortest month, which
was called the additional or intercalary day, and was
inserted in the calendar between the 23d and 24th of
that month. In the ancient Roman calendar the first
day of every month was invariably called the calends.
The 24th of February then was the 6th of the calends
of March—Sexto calendas; the preceding, which was
the additional or intercalary day, was called bis-sexto
calendas (from *bis*, twice, and *sextus*, six), twice the
sixth day. Hence the term bis-sextile as applied to
every fourth year, commonly called leap-year. Appendix B.

CHAPTER IV.

HISTORY OF THE REFORMATION OF THE JULIAN CALENDAR BY POPE GREGORY XIII.

True enough, the year in which Julius Cæsar reformed the ancient Roman calendar was the last year of confusion, and the method adopted by him a commodious one, and answered a very good purpose for a short time ; but as the years rolled on and century after century had passed away, astronomers began to discover the discrepancy between the solar and the civil year ; that the vernal equinox did not occupy the place it occupied in the time of Cæsar, namely, the 24th of March, but was gradually retrograding towards the beginning of the year, so that at the meeting of the Council of Nice in 325 it fell on the 21st. Appendix C.

The venerable Bede, in the 8th century, observed that these phenomena took place three or four days earlier than at the meeting of that council. Roger Bacon, in the 13th century, wrote a treatise on this subject and sent it to the Pope, setting forth the errors of the Julian calendar. The discrepancy at that time amounted to seven or eight days.

Thus the errors of the calendar continued to increase until 1582, when the vernal equinox fell on the 11th instead of the 21st of March. Gregory, perceiving that the measure (of reforming the calendar) was likely to confer great eclat on his pontificate, undertook the long desired reformation ; and having found the governments of the principal Catholic states ready to adopt his views, he issued a brief in the month of March, 1582, in which he abolished the use of the ancient calendar, and substituted that which has since been received in almost all Christian countries under the name of the Gregorian calendar or New Style.

The edict of the Pope took effect in October of that year, causing the 5th to be called the 15th of that month, thus suppressing ten days and making the year 1582 to consist of only 355 days. So we see that the ten days that had been gained by incorrect computation during the past 1257 years, were deducted from 1582, restoring the concurrence of the solar and the civil year, and consequently the vernal equinox to the place it occupied in 325, namely, the 21st of March.

The Pope was promptly obeyed in Spain, Portugal, and Italy. The change took place the same year in France, by calling the 10th the 20th of December. Many other Catholic countries made the change the same year, and the Catholic states of Germany the year following ; but most of the Protestant countries adhered to the Old Style until after the year 1700. Among the last was Great Britain ; she, after having suffered a great deal of inconvenience for nearly two hundred years by using a different date from the most of Europe, at length, by an act of Parliament, fixed on September, 1752, as the time for making the much desired change, which was done by calling the 3d of that month the 14th (as the error now amounted to eleven days), adopting at the same time the Gregorian rule of intercalation.

Russia is the only Christian country that still adheres to the Old Style, and by using a different date from the rest of Europe is now twelve days behind the true time. The discrepancy between solar and civil time does not effect the day, for, as has already been shown, the mean length of the day is twenty-four hours, and is marked by one revolution of the earth upon its axis.

Nor does it effect the week, for the week is uniformly seven of those days. But it effects the year, the month and the day of the month.

Russia, by adhering to the Old Style, has reckoned as many days and as many weeks, and events have transpired on the same day of the week as they have with us who have adopted the New Style ; as Christian nations we are observing the same day as the Sabbath.

When it was Tuesday, the 20th day of December, 1888, in Russia, it was Tuesday, the 1st day of January, 1889, in those countries which have adopted the New Style. Columbus sailed from Palos, in Spain, on Friday, August 3d, 1492, Old Style, which was Friday, August 12th, New Style. Washington was born on Friday, February 11th, 1732, Old Style, which was Friday, February 22d, New Style.

Now, the difference in styles during the 15th century is nine days ; during the 16th and 17th centuries, ten days ; the 18th century, eleven days, and the 19th, twelve days. In regard to the sailing of Columbus, the change is made by suppressing nine days, calling the 3d the 12th of August. In regard to the birth of Washington, the change is effected by suppressing eleven days, calling the 11th of February the 22d. As regards Russia, she could have made the change last year by calling the 20th of December, 1888, the 1st day of January, 1889, thereby suppressing twelve days, and making the year 1888 to consist of only 354 days, and the month of December twenty days. The methods of computation, both Old and New Styles, will be explained in another chapter.

To persons unacquainted with astronomy, the difference between Old and New Styles would probably be better understood by the diagram on the 25th page. The figures represent the ecliptic, which is the apparent path of the Sun, or the real path of the Earth as seen from the Sun, in her annual or yearly revolution around the Sun in the order of the months, as marked on the ecliptic.

Attention is called to four points on the ecliptic, namely, the vernal equinox, the autumnal equinox, the winter solstice, and the summer solstice. These occur, in the order given above, on the 21st of March, the 21st of September, the 21st of December and the 21st of June. It has already been stated that if the civil year correspond with the solar, the seasons of the year will always come at the same period. Julius Cæsar found the ancient Roman year in advance of the solar ; Gregory found the Julian behind the solar year; so one reforms the calendar by intercalation, the other by suppression. Appendix D.

Cæsar restored the coincidence of the solar and the civil year, but failed to retain it by allowing what probably appeared to him at the time a trifling error in his calendar. The error, which was 11 minutes and 10.38 seconds every year, was nardly perceptible for a short period, but still amounted to three days every 400 years. Hence the necessity in 1582 of reforming the reformed calendar of Julius Cæsar to restore the coincidence. Appendix E.

From the meeting of the Council of Nice, in 325, to 1582, a period of 1257 years, there was found to be an error in the Julian calendar of ten days. Now, in 1257 years the Earth performs 1257 annual and 459,109 daily revolutions, after which the vernal equinox was found to occur on the 21st of March, true or solar time ; thus concurring with the vernal equinox of 325. But the erroneous Julian calendar would make the Earth perform 459,119 daily revolutions to complete the 1257 years, a discrepancy of ten days, making the vernal equinox to fall on the 11th instead of the 21st. It will be seen by the diagram that the ten days were deducted from October, in 1582, making it a short month, consisting of only twenty-one days.

The discrepancy between the Julian and Gregorian calendar amounts to thirty days in 4000 years ; three months in 12,175 years. Hence, in 12,175 years the equinoxes would take the place of the solstices, and the solstices the place of the equinoxes. In 24,350 years, the vernal equinox would take the place of the autumnal equinox, and the winter solstice the place of the summer solstice.

And in 48,700 years, according to the Julian rule of intercalation, there would be gained nearly $365\frac{1}{4}$ days, or one entire revolution of the Earth. So, to restore the concurrence of the Julian and Gregorian years, there would have to be suppressed $365\frac{1}{4}$ days, calling the 1st day of January, 48,699, the 1st day of January, 48,700.

Thus would disappear from the Julian calendar twelve months, or one whole year, it having been divided among the thousands of the preceding years.

21st.

Jan. 1st.

W. Solstice.

Dec. 1st.

Nov. 1st.

Feb. 1st.

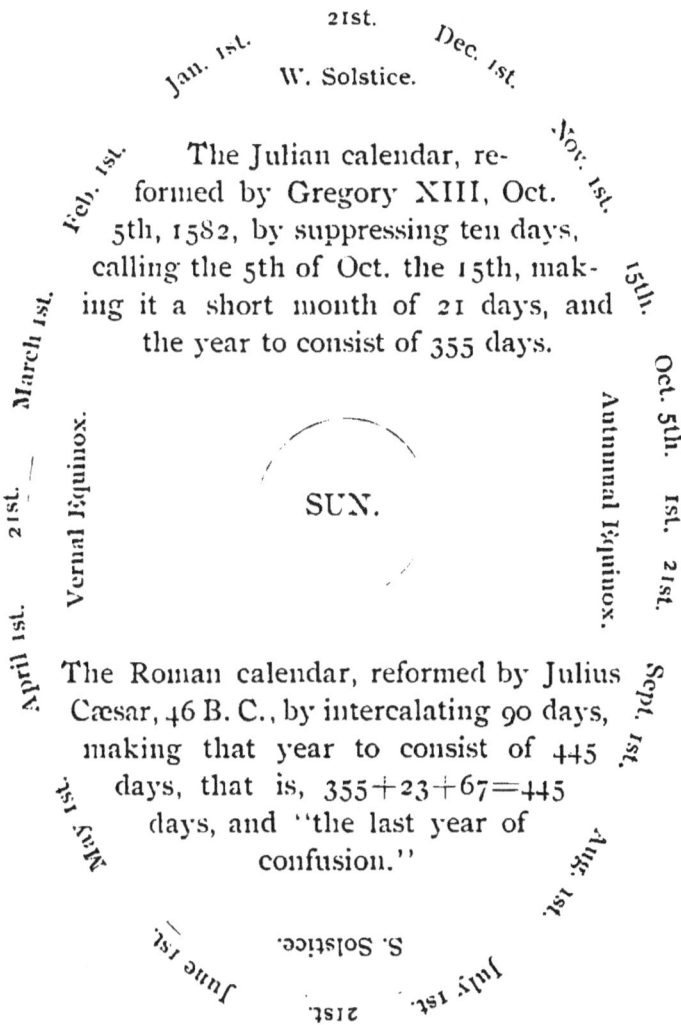

The Julian calendar, re-
formed by Gregory XIII, Oct.
5th, 1582, by suppressing ten days,
calling the 5th of Oct. the 15th, making it a short month of 21 days, and
the year to consist of 355 days.

15th.

Oct. 5th.

March 1st.

1st.

21st.

SUN.

21st.

Vernal Equinox.

Autumnal Equinox.

April 1st.

The Roman calendar, reformed by Julius
Cæsar, 46 B. C., by intercalating 90 days,
making that year to consist of 445
days, that is, $355 + 23 + 67 = 445$
days, and "the last year of
confusion."

Sept. 1st.

May 1st.

Aug. 1st.

June 1st.

S. Solstice.

July 1st.

21st.

To make this subject better understood, let us suppose the solar year to consist in round numbers of 365 days, and the civil year 366. It is evident that at the end of the year of 365 days, there would still be wanting one day to complete the civil year of 366 days, so one day must be added to that year, and to every succeeding year, to complete the years of 366 days each, which would be the loss of one year of 365 days in 365 years. Hence, 364 years of 366 days each are equal to 365 years of 365 days each, wanting one day.

Again, let us suppose the civil year to consist of 364 days. It is evident that at the end of the supposed solar year of 365 days, there would be an advance or gain of one day in that year and every succeeding year, so that in 365 years there would be a gain of 365 days or one whole year. Hence, 366 years of 364 days each are equal to 365 years of 365 days each, wanting one day. Appendix F.

CHAPTER V.

PECULIARITIES OF THE ROMAN CALENDAR.

The Romans, instead of distinguishing the days of the month by the ordinal numbers, first, second, third, etc., counted backwards from three fixed points, namely, the Calends, the Nones, and the Ides.

Calends (Latin *Calandae*, from *Calare*, to call,) was so denominated because it had been an ancient custom of the pontiffs to call the people together on that day to apprise them of the festivals, or days that were to be kept sacred during the month.

Nones (Latin *nonae*, from *nonus*, the ninth,) the ninth day before the Ides.

Ides (Latin *idus*, supposed to be derived from an obsolete verb *iduare*, to divide,) was near the middle of the month, either the 13th or the 15th day.

The first day of each month was invariably called the Calends. The Nones were the fifth, and the Ides the thirteenth, except in March, May, July, and October, in which the Nones occurred on the seventh day and the Ides on the fifteenth.

From these three points the days of the month were numbered—not forward, but backward—as so many days before the Nones, the Ides, or the Calends, the point of departure being counted in the reckoning, so that the last day of every month was the second of the Calends of the following month.

It will be seen by the Roman and English calendar found on the following pages, that there are six days of Nones in March, May, July and October, and four of all the other months; also that all the months have eight days of Ides. The number of days of Calends depend upon the number of days in the month, and the day of the month on which the Ides fall.

If the month has thirty-one days and the Ides fall on the thirteenth, there are nineteen days of Calends; but if the Ides fall on the fifteenth, there are only seventeen days of Calends. As the Ides fall on the thirteenth of all the months of thirty days, they have eighteen days of Calends. February, the month of twenty-eight days, has only sixteen, except in leap-year, when the sixth of the Calends is reckoned twice.

It may also be seen from the calendar that the

Romans, after the first day of the month, began to reckon so many days before the Nones, as 4th, 3d, 2d, then Nones ; after the Nones, so many days before the Ides, as 8th, 7th, 6th, etc., and after the Ides, so many before the Calends of the next month, the highest numbers being reckoned first.

In reducing the Roman calendar to our own, it should be remembered that in reckoning backward from a fixed point, that the point of departure is counted ; also, that the last day of the month is not the point from which the Calends are reckoned, but the first day of the following month. We have then this rule for finding the English expression for any Latin date :

RULE.

If the given date be Calends, add two to the number of days in the month, from which subtract the given date ; if the date be Nones, or Ides, add one to that of the day on which the Nones or Ides fall, from which subtract the given date, and you will have the day of the month in our calendar. To find the Latin expression for any English date, the preceding method is to be reversed, upon the principle that if $5-3=2$, then $5-2=3$.

But in reducing a Roman date to a date of February in leap-year, for the first twenty-four days, proceed according to the preceding rule as if the month had only twenty-eight days, and to obtain the proper expression for the remaining five days, regard the month as having twenty-nine days, taking the Roman date from 31 instead of 30. Thus $31-6=25$, while $30-6=24$; the former corresponding with sexto calendas, the latter with bis-sexto calendas of the Julian calendar.

By referring to the table on the 35th page, one may easily learn how to find the English expression for any Latin date, or the Latin expression for any English date.

It has already been stated that in January the Nones fall on the 5th, and the Ides on the 13th, January then having thirty one days, the number from which to subtract the Roman date to obtain the corresponding day of the month is, for Nones, $5+1=6$; for Ides, $13+1=14$; for Calends, $31+2=33$. Hence the first column in the table under January are the numbers 6, 14 and 33 February is the same as January for Nones and Ides. For Calends in leap-year, for the first 24 days it is $28+2=30$; for the remaining 5 days it is $29+2=31$. Hence for the first column under February are the numbers 6, 14, 30 and 31. March is the same as January, except that the Nones fall on the 7th and the Ides on the 15th, consequently we have for Nones, $7+1=8$, and for Ides, $15+1=16$; hence, for the first column under March we have the numbers 8, 16, and 33.

In the table the three months are taken to illustrate how easily the change may be made from Roman to English, or from English to Roman date. A complete calendar for 1892, both in Roman and English, which will be very convenient for reference, may be found on the four following pages.

The first seven letters of the alphabet, used to repre-sent the days of the week, are placed in the calendar beside the days of the week. The letter that represents Sunday is called the dominical or Sunday letter. The letter that represents the first Sunday in any given year represents all the Sundays in that year, unless it be leap-year, when two Sunday letters are used. The first represents all the Sundays in January and February, while the letter that precedes it represents all the Sundays for the rest of the year.

The reason of this is, the day intercalated, or thrust in, between the 28th day of February and the 1st day of March so interrupts the order of the letters that D, which always represents the 1st day of March, now represents the 29th day of February, so that in 1892 it represents Monday, the 29th day of February, also Tuesday the 1st day of March. As it represented all the Mondays in January and February, it will now represent all the Tuesdays for the rest of the year, while C, the letter preceding, represented all the Sundays, will now represent all the Mondays, and B all the Sundays. For January and February we have then C, Sunday ; D, Monday ; E, Tuesday ; F, Wednesday ; G, Thursday ; A, Friday, and B, Saturday. For the rest of the year we have B, Sunday ; C, Monday ; D, Tuesday ; E, Wednesday ; F, Thursday ; G, Friday, and A, Saturday. See Part Second, chapters IV and V.

JAN., 1892.			FEB., 1892.			MARCH, 1892.		
1	*Cal.*	a Fri.	1	*Cal.*	d Mon.	1	*Cal.*	d Tues.
2	4	b Sat.	2	4	e Tues.	2	6	e Wed.
3	3	c Sun.	3	3	f Wed.	3	5	f Thur.
4	2	d Mon.	4	2	g Thur.	4	4	g Fri.
5	*Non.*	e Tues.	5	*Non.*	a Fri.	5	3	a Sat.
6	8	f Wed.	6	8	b Sat	6	2	b Sun.
7	7	g Thur.	7	7	c Sun.	7	*Non*	c Mon.
8	6	a Fri.	8	6	d Mon.	8	8	d Tues.
9	5	b Sat.	9	5	e Tues.	9	7	e Wed.
10	4	c Sun.	10	4	f Wed.	10	6	f Thur.
11	3	d Mon.	11	3	g Thur.	11	5	g Fri.
12	2	e Tues.	12	2	a Fri.	12	4	a Sat.
13	*Ides.*	f Wed.	13	*Ides.*	b Sat.	13	3	b Sun.
14	19	g Thur.	14	16	c Sun.	14	2	c Mon.
15	18	a Fri.	15	15	d Mon.	15	*Ides.*	d Tues.
16	17	b Sat.	16	14	e Tues.	16	17	e Wed.
17	16	c Sun.	17	13	f Wed.	17	16	f Thur.
18	15	d Mon.	18	12	g Thur.	18	15	g Fri.
19	14	e Tues.	19	11	a Fri.	19	14	a Sat.
20	13	f Wed.	20	10	b Sat.	20	13	b Sun.
21	12	g Thur.	21	9	c Sun.	21	12	c Mon.
22	11	a Fri.	22	8	d Mon.	22	11	d Tues.
23	10	b Sat.	23	7	e Tues.	23	10	e Wed.
24	9	c Sun.	24	6	f Wed.	24	9	f Thur.
25	8	d Mon.	25	6	g Thur.	25	8	g Fri.
26	7	e Tues.	26	5	a Fri.	26	7	a Sat.
27	6	f Wed.	27	4	b Sat.	27	6	b Sun.
28	5	g Thur.	28	3	c Sun.	28	5	c Mon.
29	4	a Fri.	29	2	d Mon.	29	4	d Tues.
30	3	b Sat.				30	3	e Wed.
31	2	c Sun.				31	2	f Thur.

APRIL, 1892.				MAY, 1892.				JUNE, 1892.		
1	*Cal.*	g	Fri.	1	*Cal.*	b	Sun.	1	*Cal.*	e	Wed.
2	4	a	Sat.	2	6	c	Mon.	2	4	f	Thur.
3	3	b	Sun.	3	5	d	Tues	3	3	g	Fri.
4	2	c	Mon.	4	4	e	Wed	4	2	a	Sat.
5	*Non.*	d	Tues.	5	3	f	Thur.	5	*Non.*	b	Sun.
6	8	e	Wed.	6	2	g	Fri.	6	8	c	Mon.
7	7	f	Thur	7	*Non.*	a	Sat.	7	7	d	Tues.
8	6	g	Fri.	8	8	b	Sun	8	6	e	Wed.
9	5	a	Sat.	9	7	c	Mon.	9	5	f	Thur.
10	4	b	Sun.	10	6	d	Tues.	10	4	g	Fri.
11	3	c	Mon.	11	5	e	Wed.	11	3	a	Sat.
12	2	d	Tues.	12	4	f	Thur.	12	2	b	Sun.
13	*Ides.*	e	Wed.	13	3	g	Fri.	13	*Ides.*	c	Mon.
14	18	f	Thur.	14	2	a	Sat	14	18	d	Tues.
15	17	g	Fri.	15	*Ides.*	b	Sun.	15	17	e	Wed.
16	16	a	Sat.	16	17	c	Mon.	16	16	f	Thur.
17	15	b	Sun.	17	16	d	Tues.	17	15	g	Fri.
18	14	c	Mon.	18	15	e	Wed.	18	14	a	Sat.
19	13	d	Tues.	19	14	f	Thur.	19	13	b	Sun.
20	12	e	Wed.	20	13	g	Fri.	20	12	c	Mon.
21	11	f	Thur.	21	12	a	Sat.	21	11	d	Tues.
22	10	g	Fri.	22	11	b	Sun.	22	10	e	Wed.
23	9	a	Sat.	23	10	c	Mon.	23	9	f	Thur.
24	8	b	Sun.	24	9	d	Tues.	24	8	g	Fri.
25	7	c	Mon.	25	8	e	Wed.	25	7	a	Sat.
26	6	d	Tues.	26	7	f	Thur.	26	6	b	Sun.
27	5	e	Wed.	27	6	g	Fri.	27	5	c	Mon.
28	4	f	Thur.	28	5	a	Sat.	28	4	d	Tues.
29	3	g	Fri.	29	4	b	Sun.	29	3	e	Wed.
30	2	a	Sat.	30	3	c	Mon.	30	2	f	Thur.
				31	2	d	Tues.				

JULY, 1892.			AUG., 1892.			SEPT., 1892.		
1	*Cal.*	g Fri.	1	*Cal.*	c Mon.	1	*Cal.*	f Thur.
2	6	a Sat.	2	4	d Tues	2	4	g Fri.
3	5	b Sun.	3	3	e Wed	3	3	a Sat.
4	4	c Mon.	4	2	f Thur.	4	2	b Sun.
5	3	d Tues.	5	*Non.*	g Fri.	5	*Non.*	c Mon.
6	2	e Wed.	6	8	a Sat.	6	8	d Tues.
7	*Non.*	f Thur	7	7	b Sun.	7	7	e Wed.
8	8	g Fri.	8	6	c Mon.	8	6	f Thur.
9	7	a Sat.	9	5	d Tues.	9	5	g Fri.
10	6	b Sun	10	4	e Wed.	10	4	a Sat.
11	5	c Mon.	11	3	f Thur.	11	3	b Sun.
12	4	d Tues.	12	2	g Fri.	12	2	c Mon.
13	3	e Wed.	13	*Ides.*	a Sat.	13	*Ides.*	d Tues.
14	2	f Thur	14	19	b Sun.	14	18	e Wed.
15	*Ides.*	g Fri.	15	18	c Mon.	15	17	f Thur.
16	17	a Sat.	16	17	d Tues	16	16	g Fri.
17	16	b Sun.	17	16	e Wed.	17	15	a Sat.
18	15	c Mon.	18	15	f Thur.	18	14	b Sun.
19	14	d Tues.	19	14	g Fri.	19	13	c Mon.
20	13	e Wed.	20	13	a Sat.	20	12	d Tues.
21	12	f Thur.	21	12	b Sun.	21	11	e Wed.
22	11	g Fri.	22	11	c Mon.	22	10	f Thur.
23	10	a Sat.	23	10	d Tues.	23	9	g Fri.
24	9	b Sun.	24	9	e Wed.	24	8	a Sat.
25	8	c Mon.	25	8	f Thur.	25	7	b Sun.
26	7	d Tues.	26	7	g Fri.	26	6	c Mon.
27	6	e Wed.	27	6	a Sat.	27	5	d Tues.
28	5	f Thur	28	5	b Sun.	28	4	e Wed.
29	4	g Fri.	29	4	c Mon.	29	3	f Thur.
30	3	a Sat.	30	3	d Tues.	30	2	g Fri.
31	2	b Sun	31	2	e Wed.			

Oct., 1892.				Nov., 1892.				Dec., 1892.		
1	*Cal.*	a	Sat.	1	*Cal.*	d	Tues.	1	*Cal.*	f Thur.
2	6	b	Sun.	2	4	e	Wed.	2	4	g Fri.
3	5	.c	Mon.	3	3	f	Thur.	3	3	a Sat.
4	4	d	Tues.	4	2	g	Fri.	4	2	b Sun.
5	3	e	Wed.	5	*Non.*	a	Sat	5	*Non*	c Mon.
6	2	f	Thur.	6	8	b	Sun.	6	8	d Tues.
7	*Non.*	g	Fri.	7	7	c	Mon.	7	7	e Wed.
8	8	a	Sat.	8	6	d	Tues.	8	6	f Thur.
9	7	b	Sun.	9	5	e	Wed.	9	5	g Fri.
10	6	c	Mon.	10	4	f	Thur.	10	4	a Sat.
11	5	d	Tues.	11	3	g	Fri.	11	3	b Sun.
12	4	e	Wed.	12	2	a	Sat.	12	2	c Mon.
13	3	f	Thur.	13	*Ides.*	b	Sun.	13	*Ides.*	d Tues.
14	2	g	Fri.	14	18	c	Mon.	14	19	e Wed.
15	*Ides.*	a	Sat.	15	17	d	Tues.	15	18	f Thur.
16	17	b	Sun.	16	16	e	Wed	16	17	g Fri.
17	16	c	Mon.	17	15	f	Thur.	17	16	a Sat
18	15	d	Tues.	18	14	g	Fri.	18	15	b Sun.
19	14	e	Wed.	19	13	a	Sat.	19	14	c Mon.
20	13	f	Thur.	20	12	b	Sun.	20	13	d Tues.
21	12	g	Fri.	21	11	c	Mon.	21	12	e Wed.
22	11	a	Sat.	22	10	d	Tues.	22	11	f Thur.
23	10	b	Sun.	23	9	e	Wed.	23	10	g Fri.
24	9	c	Mon.	24	8	f	Thur.	24	9	a Sat.
25	8	d	Tues.	25	7	g	Fri.	25	8	b Sun.
26	7	e	Wed.	26	6	a	Sat.	26	7	c Mon.
27	6	f	Thur.	27	5	b	Sun.	27	6	d Tues.
28	5	g	Fri.	28	4	c	Mon.	28	5	e Wed.
29	4	a	Sat.	29	3	d	Tues.	29	4	f Thur.
30	3	b	Sun.	30	2	e	Wed.	30	3	g Fri.
31	2	c	Mon.					31	2	a Sat.

JANUARY.	FEBRUARY.	MARCH.
Cal. 1	*Cal.* 1	*Cal.* 1
6— 4= 2	6— 4= 2	8— 6= 2
6— 3= 3	6— 3= 3	8— 5= 3
6— 2= 4	6— 2= 4	8— 4= 4
Nones 5	*Nones* 5	8— 3= 5
14— 8= 6	14— 8= 6	8— 2= 6
14— 7= 7	14— 7= 7	*Nones.* 7
14— 6= 8	14— 6= 8	16— 8= 8
14— 5= 9	14— 5= 9	16— 7= 9
14— 4=10	14— 4=10	16— 6=10
14— 3=11	14— 3=11	16— 5=11
14— 2=12	14— 2=12	16— 4=12
Ides 13	*Ides* 13	16— 3=13
33—19=14	30—16=14	16— 2=14
33—18=15	30—15=15	*Ides* 15
33—17=16	30—14=16	33—17=16
33—16=17	30—13=17	33—16=17
33—15=18	30—12=18	33—15=18
33—14=19	30—11=19	33—14=19
33—13=20	30—10=20	33—13=20
33—12=21	30— 9=21	33—12=21
33—11=22	30— 8=22	33—11=22
33—10=23	30— 7=23	33—10=23
33— 9=24	30— 6=24	33— 9=24
33— 8=25	31— 6=25	33— 8=25
33— 7=26	31— 5=26	33— 7=26
33— 6=27	31— 4=27	33— 6=27
33— 5=28	31— 3=28	33— 5=28
33— 4=29	31— 2=29	33— 4=29
33— 3=30		33— 3=30
33— 2=31		33— 2=31

36

PART SECOND.

MATHEMATICAL.

CHAPTER I.

ERRORS OF THE JULIAN CALENDAR.

It will be necessary in the first place to understand the difference between the Julian and Gregorian rule of intercalation. If the number of any year be exactly divisible by four it is leap year; if the remainder be 1, it is the first year after leap-year; if 2, the second; if 3, the third; thus:

$$1888 \div 4 = 472, \text{ no remainder.}$$
$$1889 \div 4 = 472, \text{ remainder, } 1.$$
$$1890 \div 4 = 472, \text{ remainder, } 2.$$
$$1891 \div 4 = 472, \text{ remainder, } 3.$$
$$1892 \div 4 = 473, \text{ no remainder.}$$

And so on, every fourth year being leap-year of 366 days.

This is the Julian rule of intercalation, which is corrected by the Gregorian by making every centurial year, or the year that completes the century, a common year, if not exactly divisible by 400; so that only every fourth centurial year is leap-year; thus, 1,700, 1,800, and 1,900 are common years, but 2,000, the fourth centurial year, is leap year, and so on.

By the Julian rule three-fourths of a day is gained every century, which in 400 years amounts to three days. This is corrected by the Gregorian, by making three consecutive centurial years common years, thus suppressing three days in 400 years.

RULE.

Multiply the difference between the Julian and the solar year by 100, and we have the error in 100 years. Multiply this product by 4 and we have the error in 400 years. Now, 400 is the tenth of 4,000; therefore, multiply the last product by 10, and we have the error in 4,000 years. Now, as the discrepancy between the Julian and Gregorian year is three days in 400 years, making 3-400 of a day every year, so by dividing $365\frac{1}{4}$, the number of days in a year, by 3-400, we have the time it would take to make a revolution of the seasons.

SOLUTION.

(365 d, 6 h.)—(365 d, 5 h, 48 m, 49.62 s.)$=$(11 m, 10.38 s.) Now, (11 m, 10.38 s.)×100$=$18 h, 37.3 m, the gain in 100 years. This is, reckoned in round numbers, 18 hours, or three-fourths of a day. Now, $(\frac{3}{4}\times4)=(1\times3)=3$: the Julian rule gaining three days, the Gregorian suppressing three days in 400 years. $(3\times10)=30$, the number of days gained by the Julian rule in 4 000 years. $365\frac{1}{4}\div3\,400=48,700$, so that in this long period of time, this falling back $\frac{3}{4}$ of a day every century would amount to $365\frac{1}{4}$ days; therefore, 48,699 Julian years are equal to 48,700 Gregorian years.

CHAPTER II.

ERRORS OF THE GREGORIAN CALENDAR.

By reference to the preceding chapter it will be seen that there is an error of 37.3 minutes in every 100 years not corrected by the Gregorian calendar; this amounts to only .373 of a minute a year, or one day in 3,861 years, and one day and fifty-two minutes in 4,000 years.

RULE.

To find how long it would take to gain one day: Divide the number of minutes in a day by the decimal .373, that being the fraction of a minute gained every year. To find how much time would be gained in 4,000 years, multiply the decimal .373 by 4,000, and you will have the answer in minutes, which must be reduced to hours.

SOLUTION.

$(24 \times 60) \div .373 = 3,861$, nearly; hence the error would amount to only one day in 3,861 years.

$(.373 \times 4,000) \div 60 = (24 \text{ h}, 52 \text{ m},) = (1 \text{ d}, 0 \text{ h}, 52 \text{ m})$, the error in 4,000 years.

This trifling error in the Gregorian calendar may be corrected by suppressing the intercalations in the year 4,000, and its multiples, 8,000, 12,000, 16,000, etc., so that it will not amount to a day in 100,000 years.

RULE.

Divide 100,000 by 4,000 and you will have the number of intercalations suppressed in 100,000 years. Multiply 1 d, 52 m, (that being the error in 4,000 years) by this quotient, and you will have the discrepancy between the Gregorian and solar year for 100,000 years. By this improved method we suppress 25 days, so that the error will only amount to 25 times 52 minutes.

SOLUTION.

100,000÷4,000×(1 d, 52 m,)=(25d, 21 h, 40 m.) Now, (25d, 21 h, 40 m,)—25 d=(21 h, 40 m,) the error in 100,000.

CHAPTER III.

DOMINICAL LETTER.

Dominical (from the Latin *Dominus*, Lord,) indicating the Lord's day or Sunday. Dominical letter, one of the first seven letters of the alphabet used to denote the Sabbath or Lord's day.

For the sake of greater generality, the days of the week are denoted by the first seven letters of the alphabet, A, B, C, D, E, F, G, which are placed in the calendar beside the days of the year, so that A stands opposite the first day of January, B opposite the second, C opposite the third, and so on to G, which stands opposite the seventh ; after which A returns to the eighth, and so on through the 365 days of the year.

Now, if one of the days of the week, Sunday for example, is represented by F, Monday will be represented by G, Tuesday by A, Wednesday by B, Thursday by C, Friday by D, and Saturday by E; and every Sunday throughout the year will have the same character, F, every Monday G, every Tuesday A, and so with regard to the rest.

The letter which denotes Sunday is called the Dominical or Sunday letter for that year; and when the dominical letter of the year is known, the letters which respectively correspond to the other days of the week become known also. Did the year consist of 364 days, or 52 weeks invariably, the first day of the year and the first day of the month, and in fact any day of any year, or any month, would always commence on the same day of the week. But every common year consists of 365 days, or 52 weeks and 1 day, so that the following year will begin one day later in the week than the year preceding. Thus the year 1837 commenced on Sunday, the following year, 1838, on Monday, 1839 on Tuesday, and so on.

As the year consists of 52 weeks and 1 day, it is evident that the day which begins and ends the year must occur 53 times; thus the year 1837 begins on Sunday and ends on Sunday; so the following year, 1838, must begin on Monday. As A represented all the Sundays in 1837, and as A always stands for the first day of January, so in 1838 it will represent all the Mondays, and the dominical letter goes back from A to G; so that G represents all the Sundays in 1838, A all the Mondays, B all the Tuesdays, and so on, the dominical letter going back one place in every year of 365 days.

While the following year commences one day later in the week than the year preceding, the dominical letter goes back one place from the preceding year; thus while the year 1865 commenced on Sunday, 1866 on Monday, 1867 on Tuesday, the dominical letters are A, G and F, respectively. Therefore, if every year con-

sisted of 365 days, the dominical cycle would be completed in seven years, so that after seven years the first
day of the year would again occur on the same day of
the week.

But this order is interrupted every four years by
giving February 29 days, thereby making the year to
consist of 366 days, which is 52 weeks and two days,
so that the following year would commence two days
later in the week than the year preceding, thus the
year 1888 being leap-year, had two dominical letters,
A and G ; A for January and February, and G for
the rest of the year. The year commenced on Sunday
and ended on Monday, making 53 Sundays and 53
Mondays, and the following year, 1889, to commence
on Tuesday. It now becomes evident that if the years
all consisted of 364 days, or 52 weeks, they would all
commence on the same day of the week ; if they all
consisted of 365 days, or 52 weeks and one day, they
would all commence one day later in the week than
the year preceding ; if they all consisted of 366 days,
or 52 weeks and two days, they would commence two
days later in the week ; if 367 days or 52 weeks and
three days, then three days later, and so on, one day
later for every additional day. It is also evident that
every additional day causes the dominical letter to go
back one place. Now in leap-year the 29th day of
February is the additional or intercalary day. So one
letter for January and February, and another for the
rest of the year. If the number of years in the intercalary period were two, and seven being the number of
days in the week, their product would be $2 \times 7 = 14$;
fourteen, then, would be the number of years in the

cycle.   Again, if the number of years in the intercalary period were three, and the number of days in the
week being seven, their product would be $3 \times 7 = 21$;
twenty-one would then be the number of years in the
cycle.   But the number of years in the intercalary
period is four, and the number of days in the week is
seven, therefore their product is $4 \times 7 = 28$;   twenty-
eight is then the number of years in the cycle.

This period is called the dominical or solar cycle,
and restores the first day of the year to the same day
of the week.   At the end of the cycle the dominical
letters return again in the same order, on the same
days of the month.   Thus, for the year 1801, the dominical letter is D; 1802, C; 1803, B; 1804, A and G;
and so on, going back five places every four years for
twenty-eight years, when the cycle, being ended, D is
again dominical letter for 1829, C for 1830, and so on
every 28 years forever, according to the Julian rule of
intercalation.

But this order is interrupted in the Gregorian calendar at the end of the century by the secular suppression
of the leap-year.   It is not interrupted, however. at
the end of every century, for the leap-year is not suppressed in every fourth centurial year ;   consequently
the cycle will then be continued for two hundred years.
It should be here stated that this order continued without interruption from the commencement of the era
until the reformation of the calendar in 1582, during
which time the Julian calendar, or Old Style was used.

It has already been shown that if the number of
years in the intercalary period be multiplied by seven,
the number of days in the week, their product will be

the number of years in the cycle. Now, in the Gregorian calendar, the intercalary period is 400 years; this number being multiplied by seven, their product would be 2,800 years, as the interval in which the coincidence is restored between the days of the year and the days of the week.

This long period, however, may be reduced to 400 years; for since the dominical letter goes back five places every four years, in 400 years it will go back 500 places in the Julian and 497 in the Gregorian calendar, three intercalations being suppressed in the Gregorian every 400 years. Now 497 is exactly divisible by seven, the number of days in the week, therefore, after 400 years the cycle will be completed, and the dominical letters will return again in the same order, on the same days of the month.

In answer to the question, "Why two dominical letters for leap-year?' we reply, because of the additional or intercalary day after the 28th of February. It has already been shown that every additional day causes the dominical letter to go back one place. As there are 366 days in leap-year, the letter must go back two places, one being used for January and February, and the other for the rest of the year. Did we continue one letter through the year and then go back two places, it would cause confusion in computation, unless the intercalation be made at the end of the year. Whenever the intercalation is made there must necessarily be a change in the dominical letter. Had it been so arranged that the additional day was placed after the 30th of June or September, then the first letter would be used until the intercalation is made in June or

September, and the second to the end of the year. Or suppose that the end of the year had been fixed as the time and place for the intercalation, (which would have been much more convenient for computation,) then there would have been no use whatever for the second dominical letter, but at the end of the year we would go back two places ; thus, in the year 1888, instead of A being dominical letter for two months merely, it would be continued through the year, and then passing back to F, no use whatever being made of G, and so on at the end of every leap-year. Hence it is evident that this arrangement would have been much more convenient, but we have this order of the months, and the number of days in the months as Augustus Cæsar left them eight years before Christ. The dominical letter probably was not known until the Council of Nice, in the year of our Lord 325, where, in all probability, it had its origin.

CHAPTER IV.

RULE FOR FINDING THE DOMINICAL LETTER.

Divide the number of the given year by 4, neglecting the remainders, and add the quotient to the given number. Divide this amount by 7, and if the remainder be less than three, take it from 3 ; but if it be 3 or more than 3, take it from 10 and the remainder will be the number of the letter calling A, 1; B, 2; C, 3, etc.

By this rule the dominical letter is found from the commencement of the era to October 5th, 1582. O. S.

From October 15th, 1582, till the year 1700, take the remainder as found by the rule from 6, if it be less than 6, but if the remainder be 6, take it from 13, and so on according to instructions given in the table on 49th page. It should be understood here, that in leap-year the letter found by the preceding rule will be dominical letter for that part of the year that follows the 29th of February, while the letter which follows it will be the one for January and February.

EXAMPLES.

To find the dominical letter for 1365, we have $1365 \div 4 = 341 +$; $1365 + 341 = 1706$; $1706 \div 7 = 243$, remainder 5. Then $10 - 5 = 5$; therefore E being the fifth letter is the dominical letter for 1365.

To find the dominical letter for 1620, we have $1620 \div 4 = 405$; $1620 + 405 = 2025$; $2025 \div 7 = 289$, remainder 2. Then $6 - 2 = 4$; therefore, D and E are the dominical letters for 1620; E for January and February, and D for the rest of the year. The process of finding the dominical letter is very simple and easily understood, if we observe the following order:

1st. Divide by 4.

2d. Add to the given number.

3d. Divide by 7.

4th. Take the remainder from 3 or 10, from the commencement of the era to October 5th, 1582. From October 15th, 1582 to 1700, from 6 or 13. From 1700 to 1800, from 7, and so on. See table on 49th page.

We divide by 4 because the intercalary period is four years; and as every fourth year contains the divisor 4 once more than any of the three preceding years, so there is one more added to the fourth year than there is to any of the three preceding years; and as every year consists of 52 weeks and one day, this additional year gives an additional day to the remainder after dividing by 7. For example, the year

1 of the era consists of	52 w. 1 d.
2 years consist of	104 w. 2 d.
3 years consist of	156 w. 3 d.
$(4 \div 4) + 4 = 5$ years consist of	260 w. 5 d.

Hence the numbers thus formed will be 1, 2, 3, 5, 6, 7, 8, 10, 11, 12, 13, 15, and so on.

We divide by 7, because there are seven days in the week, and the remainders show how many days more than an even number of weeks there are in the given year. Take, for example, the first twelve years of the era after being increased by one-fourth, and we have

$$1 \div 7 = 0 \text{ remainder } 1 \quad \text{Then} \quad 3 - 1 = 2 = B$$
$$2 \div 7 = 0 \quad `` \quad 2 \quad `` \quad 3 - 2 = 1 = A$$
$$3 \div 7 = 0 \quad `` \quad 3 \quad `` \quad 10 - 3 = 7 = G$$
$$5 \div 7 = 0 \quad `` \quad 5 \quad `` \quad 10 - 5 = 5 = F \; E$$
$$6 \div 7 = 0 \quad `` \quad 6 \quad `` \quad 10 - 6 = 4 = D$$
$$7 \div 7 = 1 \quad `` \quad 0 \quad `` \quad 3 - 0 = 3 = C$$
$$8 \div 7 = 1 \quad `` \quad 1 \quad `` \quad 3 - 1 = 2 = B$$
$$10 \div 7 = 1 \quad `` \quad 3 \quad `` \quad 10 - 3 = 7 = A \; G$$
$$11 \div 7 = 1 \quad `` \quad 4 \quad `` \quad 10 - 4 = 6 = F$$
$$12 \div 7 = 1 \quad `` \quad 5 \quad `` \quad 10 - 5 = 5 = E$$
$$13 \div 7 = 1 \quad `` \quad 6 \quad `` \quad 10 - 6 = 4 = D$$
$$15 \div 7 = 2 \quad `` \quad 1 \quad `` \quad 3 - 1 = 2 = C \; B$$

From this table it may be seen that it is these remainders representing the number of days more than an even number of weeks in the given year, that we have to deal with in finding the dominical letter.

Did the year consist of 364 days, or 52 weeks, invariably, there would be no change in the dominical letter from year to year, but the letter that represents Sunday in any given year would represent Sunday in every year. Did the year consist of only 363 days, thus wanting one day of an even number of weeks, then these remainders, instead of being taken from a given remainder, would be added to that number, thus removing the dominical letter forward one place, and the beginning of the year, instead of being one day later, would be one day earlier in the week than in the preceding year.

Thus, if the year 1 of the era be taken from 3, we would have $3-1=2$; therefore, B being the second letter, is dominical letter for the year 1. But if the year consist of only 363 days, then the 1 instead of being taken from 3 would be added to 3; then we would have $3+1=4$; therefore, D being the fourth letter would be dominical letter for the year 1. The former going back from C to B, the latter forward from C to D; or which amounts to the same thing, make the year to consist of 51 weeks and 6 days; then $10-6=4$, making D the dominical letter as before.

As seven is the number of days in the week, and the object of these subtractions is to remove the dominical letter back one place every common year, and two in leap-year, why not take these remainders from 7? We answer, all depends upon the day of the week on which the era commenced. Had G, the seventh letter been dominical letter for the year preceding the era, then these remainders would be taken from 7; and 7 would be used until change of style in 1582. But we know from computation that C, the third letter, is dominical

letter for the year preceding the era ; so we commence
with three, and take the smaller remainders, 1 and 2
from 3 ; that brings us to A. We take the larger remainders, from 3 to 6, from 3+7=10. We add the 7
because there are seven days in the week. We use the
number 10 until we get back to C, the third letter, the
place from whence we started. For example, we have

$$3-1=2=B$$
$$3-2=1=A$$
$$10-3=7=G$$
$$10-4=6=F$$
$$10-5=5=E$$
$$10-6=4=D$$
$$3-0=3=C$$

The cycle of seven days being completed, we commence with the number three again, and so on until
1582, when on account of the errors of the Julian calendar, ten days were suppressed to restore the coincidence of the solar and civil year. Now every day suppressed removes the dominical letter forward one place;
so counting from C to C again is seven, D is eight, E
is nine, and F is ten. As F is the sixth letter, we take
the remainders from 1 to 5, from 6 ; if the remainder
be 6, take it from 6+7=13. Then 6 or 13 is used till
1700, when, another day being suppressed, the number
is increased to 7. And again in 1800, for the same
reason, a change is made to 1 or 8 ; in 1900 to 2 or 9,
and so on. It will be seen by the table on the 49th
page that the smaller numbers run from 1 to 7 ; the
larger ones from 8 to 13.

From the commencement of the Christian era to October 5th, 1582, take the remainders, after dividing by 7, from 3 or 10 ; from October 15th,

1582 to 1700 . . . . from	6 or	13
1700 to 1800 . . . . "	7	
1800 to 1900 . . . . "	1 or	8
1900 to 2100 . . . . "	2 or	9
2100 to 2200 . . . . "	3 or	10
2200 to 2300 . . . . "	4 or	11
2300 to 2500 . . . . "	5 or	12
2500 to 2600 . . . . "	6 or	13
2600 to 2700 . . . . "	7	
2700 to 2900 . . . . "	1 or	8
2900 to 3000 . . . . "	2 or	9
3000 to 3100 . . . . "	3 or	10
3100 to 3300 . . . . "	4 or	11
3300 to 3400 . . . . "	5 or	12
3400 to 3500 . . . . "	6 or	13
3500 to 3700 . . . . "	7	
3700 to 3800 . . . . "	1 or	8
3800 to 3900 . . . . "	2 or	9
3900 to 4000 . . . . "	3 or	10
4000 to 4100 . . . . "	4 or	11
4100 to 4200 . . . . "	5 or	12
4200 to 4300 . . . . "	6 or	13
4300 to 4500 . . . . "	7	
4500 to 4600 . . . . "	1 or	8
4600 to 4700 . . . . "	2 or	9
4700 to 4900 . . . . "	3 or	10
4900 to 5000 . . . . "	4 or	11
5000 to 5100 . . . . "	5 or	12

CHAPTER V.

RULE FOR FINDING THE DAY OF THE WEEK OF ANY GIVEN DATE, FOR BOTH OLD AND NEW STYLES.

By arranging the dominical letters in the order in which the different months commence, the day of the week on which any month of any year, or day of the month has fallen or will fall, from the commencement of the Christian era to the year of our Lord 4000, may be calculated. (Appendix G.) They have been arranged thus in the following couplet, in which At stands for January, Dover for February, Dwells for March, etc.

> At Dover Dwells George Brown, Esquire,
> Good Carlos Finch, and David Fryer.

Now if A be dominical or Sunday letter for a given year, then January and October being represented by the same letter, begin on Sunday ; February, March and November, for the same reason, begin on Wednesday ; April and July on Saturday ; May on Monday, June on Thursday, August on Tuesday, September and December on Friday. It is evident that every month in the year must commence on some one day of the week represented by one of the first seven letters of the alphabet. Now let

January 1st be represented   by A, Sun.
Feb. 1st (4 w. 3 d. from the preceding date) by D, Wed.
Mar. 1st 4 w. 0 d.   "   "   "   by D, Wed.
Apr. 1st 4 w. 3 d.   "   "   "   by G, Sat.
May 1st 4 w. 2 d.   "   "   "   by B, Mon.
June 1st 4 w. 3 d.   "   "   "   by E, Thur.

July 1st 4 w. 2 d. from the preceding date by G, Sat.
Aug. 1st 4 w. 3 d. " " " by C, Tues.
Sept. 1st 4 w. 3 d. " " " by F, Fri.
Oct. 1st 4 w. 2 d. " " " by A, Sun.
Nov. 1st 4 w. 3 d. " " " by D, Wed.
Dec. 1st 4 w. 2 d. " " " by F, Fri.

Now each of these letters placed opposite the months respectively represents the day of the week on which the month commences, and they are the first letters of each word in the preceding couplet.

To find the day of the week on which a given day of any year will occur, we have the following

RULE.

Find the dominical letter for the year. Read from this to the letter which begins the given month, always reading from A toward G, calling the dominical letter Sunday, the next Monday, etc. This will show on what day of the week the month commenced ; then reckoning the number of days from this will give the day required.

EXAMPLES.

History records the fall of Constantinople on May 29th, 1453. On what day of the week did it occur? We have then $1453 \div 4 = 363+$; $1453 + 363 = 1816$; $1816 \div 7 = 259$, remainder 3. Then $10 - 3 = 7$; therefore, G being the seventh letter is dominical letter for 1453. Now reading from G to B, the letter for May, we have G Sunday, A Monday, and B Tuesday; hence May commenced on Tuesday and the 29th was Tuesday.

The change from Old to New Style was made by Pope Gregory XIII, October 5th, 1582. On what day

of the week did it occur ?   We have then $1582 \div 4$-
$395+$; $1582+395=1977$; $1977 \div 7=282$, remainder 3.
Then $10-3=7$; therefore, G being the seventh letter,
is dominical letter for 1582.   Now reading from G to
A, the letter for October, we have G Sunday, A Monday, etc.   Hence October commenced on Monday,
and the 5th was Friday.

On what day of the week did the 15th of the same
month fall in 1582 ?   We have then $1582 \div 4 = 395+$;
$1582 + 395 = 1977$; $1977 \div 7 = 282$, remainder 3.   Then
$6 - 3 = 3$; therefore, C being the third letter, is the
dominical letter for 1582.   Now reading from C to A,
the letter for October, we have C Sunday, D Monday,
E Tuesday, etc.   Hence October commenced on Friday, and the 15th was Friday.

How is this, says one ?   You have just shown by
computation that October, 1582, commenced on Monday, you now say that it occurred on Friday.   You
also stated that the 5th was Friday ; you now say that
the 15th was Friday.   This is absurd ; ten is not a
multiple of seven.   There is nothing absurd about it.
The former computation was Old Style, the latter New
Style, the Old being ten days behind the new.

As regards an interval of ten days between the two
Fridays, there was none ;   Friday, the 5th, and Friday,
the 15th, was one and the same day ; there was no
interval, nothing ever occurred, there was no time for
anything to occur ; the edict of the Pope decided it ;
he said the 5th should be called the 15th, and it was so.

Hence to October the 5th, 1582, the computation
should be Old Style ; from the 15th to the end of the
year New Style.

On what day of the week did the years 1, 2 and 3, of the era commence? None of these numbers can be divided by 4; neither are they divisible by 7; but they may be treated as remainders after dividing by 7. Now each of these numbers of years consists of an even number of weeks with remainders of 1, 2 and 3 days respectively. Hence we have then for the year 1, $3-1=2$; therefore, B being the second letter, is the dominical letter for the year 1. Now reading from B to A, the letter for January, we have B Sunday, C Monday, D Tuesday, etc. Hence January commenced on Saturday.

Then we have for the year 2. $3-2=1$; therefore A being the first letter, is dominical letter for the year 2; hence it is evident that January commenced on Sunday. Again we have for the year 3, $10-3$ 7; therefore, G being the seventh letter, is dominical letter for the year 3. Now reading from G to A, the letter for January, we have G Sunday, A Monday; hence January commenced on Monday.

On what day of the week did the year 4 commence? Now we have a number that is divisible by 4, it being the first leap-year in the era, so we have $4 \div 4 = 1$; $4 + 1 = 5$; $5 \div 7 = 0$, remainder 5. Then $10 - 5 = 5$; therefore, E being the 5th letter, is dominical letter for that part of the year which follows the 29th of February, while F, the letter that follows it, is dominical letter for January and February. Now reading from F to A, the letter for January, we have F Sunday, G Monday, A Tuesday; hence January commenced on Tuesday.

Now we have disposed of the first four years of the

era; the dominical letters being B, A, G, and F, E. Hence it is evident, while one year consists of an even number of weeks and one day, two years of an even number of weeks and two days, three years of an even number of weeks and three days, that every fourth year, by intercalation, is made to consist of 366 days; so that four years consist of an even number of weeks and five days; for we have $(4 \div 4) + 4 = 5$, the dominical letter going back from G in the year 3, to F, for January and February in the year 4, and from F to E for the rest of the year, causing the following year to commence two days later in the week than the year preceding.

The year 1 had 53 Saturdays; the year 2, 53 Sundays; the year 3, 53 Mondays, and the year 4, 53 Tuesdays and 53 Wednesdays, causing the year 5 to commence on Thursday, two days later in the week than the preceding year Now what is true concerning the first four years of the era, is true concerning all the future years, and the reason for the divisions, additions and subtractions in finding the dominical letter is evident.

The Declaration of Independence was signed July 4, 1776. On what day of the week did it occur? We have then $1776 \div 4 = 444$; $1776 + 444 = 2220$; $2220 \div 7 = 317$, remainder 1. Then $7 - 1 = 6$, therefore F and G are the dominical letters for 1776, G for January and February, and F for the rest of the year. Now reading from F to G, the letter for July, we have F Sunday, G Monday; hence July commenced on Monday, and the fourth was Thursday. On what day of the week did Lee surrender to Grant, which occurred

on April 9th, 1865? We have then $1865 \div 4 = 466 +$;
$1865 - 466 = 2331$; $2331 \div 7 = 333$, remainder 0. Then
$1 - 0 = 1$; therefore, A being the first letter, is dominical
letter for 1865. Now reading from A to G, the letter
for April, we have A Sunday, B Monday, C Tuesday,
etc. Hence April commenced on Saturday, and the
9th was Sunday.

Benjamin Harrison was inaugurated President of the
United States on Monday, March 4, 1889. On what
day of the week will the 4th of March fall in 1989?
We have then $1989 \div 4 = 497 +$; $1989 + 497 = 2486$;
$2486 \div 7 = 355$, remainder 1. Then $2 - 1 = 1$, therefore,
A being the first letter, is dominical letter for 1989.
Now, reading from A to D, the letter for March, we
have A Sunday, B Monday, C Tuesday, and D Wednesday; hence March will commence on Wednesday,
and the 4th will fall on Saturday. Columbus landed
on the island of San Salvador on Friday, October 12,
1492. On what day of the month and on what day of
the week will the four hundredth anniversary fall in
1892?

The day of the month on which Columbus landed is,
of course, the day to be observed in commemoration of
that event. The Julian calendar, which was then in
use throughout Europe, and the very best that had
ever been given to the world, made the year too long
by more than eleven minutes. Those eleven minutes
a year had accumulated, from the council of Nice, in
325, to the discovery of America, in 1492, to nine days,
so that the civil year was nine days behind the true or
solar time; that is, when the Earth, in her annual
revolution, had arrived at that point of the ecliptic

coinciding with the 21st of October, the civil year, according to the Julian calendar, was the 12th.

Now, to restore the coincidence, the nine days must be dropped, or suppressed, calling what was erroneously called the 12th of October, the 21st. Since the Julian calendar was corrected by Gregory, in 1582, we have so intercalated as to retain, very nearly, the coincidence of the solar and the civil year. It has already been shown in Chapter III, (q. v.) that in the Gregorian calendar, the cycle which restores the coincidence of the day of the month and the day of the week, is completed in 400 years; so that after 400 years, events will again transpire in the same order, on the same day of the week. Now, as Columbus landed on Friday, October 21st, 1492, so Friday, October 21st, 1892, is the day of the month and also the day of the week to be observed in commemoration of that event. We have then $1892 \div 4 = 473$; $1892 + 473 = 2365$; $2365 \div 7 = 337$, remainder 6. Then $8 - 6 = 2$; therefore, B and C are dominical letters for 1892, C for January and February, and B for the rest of the year. Now, reading from B to A, the letter for October, we have B Sunday, C Monday, etc. Hence October will commence on Saturday and the 21st will be Friday.

Although there was an error of thirteen days in the Julian calendar when it was reformed by Gregory, in 1582, there was a correction made of only ten days. There was still an error of three days from the time of Julius Cæsar to the Council of Nice, which remained uncorrected. Gregory restored the vernal equinox to the 21st of March, its date at the meeting of that council, not to the place it occupied in the time of

Cæsar, namely, the 24th of March. Had he done so it would now fall on the 24th, by adopting the Gregorian rule of intercalation. Appendix H.

If desirable calculations may be made in both Old and New Styles from the year of our Lord 300. There is no perceptible discrepancy in the calendars, however, until the close of the 4th century, when it amounts to nearly one day, reckoned in round numbers one day. Now in order to make the calculation, proceed according to rule already given for finding the dominical letter, and for New Style take the remainders after dividing by seven from the numbers in the following table :

From	400	to	500		From	4	or	11
"	500	"	600		"	5	"	12
"	600	"	700		"	6	"	13
"	700	"	900		"	7		
"	900	"	1000		"	1	"	8
"	1000	"	1100		"	2	"	9
"	1100	"	1300		"	3	"	10
"	1300	"	1400		"	4	"	11
"	1400	"	1500		"	5	"	12
"	1500	"	1700		"	6	"	13

It will be found by calculation that from the year

400	to	500	the discrepancy is			1	day
500	"	600	"	"	"	2	"
600	"	700	"	"	"	3	"
700	"	900	"	"	"	4	"
900	"	1000	"	"	"	5	"
1000	"	1100	"	"	"	6	"
1100	"	1300	"	"	"	7	"
1300	"	1400	"	"	"	8	"
1400	"	1500	"	"	"	9	"
1500	"	1700	"	"	"	10	"

Hence the necessity, in reforming the calendar in 1582, of suppressing ten days. (See table on 59th page.) On what day of the week did January commence in 450? We have then $450 \div 4 = 112 +$; $450 \div 112 = 562$; $562 \div 7 = 80$, remainder 2. Then $3 - 2 = 1$; therefore, A being the first letter, is dominical letter for 450, Old Style, and January commenced on Sunday. For New Style we have $4 - 2 = 2$; Therefore, B being the second letter, is dominical letter for the year 450. Now reading from B to A, the letter for January, we have B Sunday, C Monday, D Tuesday, etc.

Hence, January commenced on Saturday. Old Style makes Sunday the first day ; New Style makes Saturday the first and Sunday the second. On what day of the week did January commence in the year 1250? We have then $1250 \div 4 = 312 +$; $1250 + 312 = 1562$; $1562 \div 7 = 223$, remainder 1. Then $3 - 1 = 2$; therefore, B being the second letter, is dominical letter for the year 1250, Old Style. Now, reading from B to A, the letter for January, we have B Sunday, C Monday, etc. Hence January commenced on Saturday. B is also dominical letter, New Style; for we take the remainder after dividing by 7, from the same number.

As both Old and New Styles have the same dominical letter, so both make January to commence on the same day of the week ; but Old Style, during this century, is seven days behind the true time, so that when it is the first day of January by the Old, it is the eighth by the New.

Vernal equinox in the time of Numa, about 700 B. C. . . . .

It is here seen by the errors of the Julian Calen-
    dar the Vernal Equinox is made to occur
    three days earlier every 400 years, so
    that in 1582 it fell on the
    11th instead of the
    21st of March.   18

                                17 . . . . .

                        16 . . . .

                  15 . . . .

              14 . . . .

          13 . . . .

      12 . . . .

11 . . . . By suppressing 1

                                     F

                                     F

By the Gregorian rule of intercalation the coincidence
years.

		March 24,	46 B. C.
		23 . . .	1 A. D.
		22	100 " "
		21	300 " "
		20 . . .	400 " "
		19	500 " "
			600 " "
			800 " "
			900 " "
			1000 " "
			1200 " "
			1300 " "
			1400 " "

Vernal equinox at the Council of Nice, 325 A. D.

Restored by Julius Cæsar, 46 B. C., to the place it occupied in the time of Numa.

.ys, Coincidence	0	Restored in	1600 " "
s behind time,	18	6 in advance.	1700 " "
" "	12	12 " "	1800 " "
" "	6	18 " "	1900 " "
Coincidence	0	Restored.	2000 " "
s behind time,	18	6 in advance.	2100 " "
" "	12	12 " "	2200 " "
" "	6	18 " "	2300 " "
Coincidence	0	Restored.	2400 " "

:he solar and civil year is restored very nearly every 400

Appendix I.

CHAPTER VI.

A SIMPLE METHOD FOR FINDING THE DAY OF THE WEEK OF EVENTS WHICH OCCUR QUADRENNIALLY.

The inaugural of the Presidents. The day of the week on which they have occurred, and on which they will occur for the next one hundred years :

April 30th,	1789,	Thursday,	George Washington.
March 4th,	1793,	Monday,	" "
" "	1797,	Saturday,	John Adams.
" "	1801,	Wednesday,	Thomas Jefferson.
" "	1805,	Monday,	" "
" "	1809,	Saturday,	James Madison.
" "	1813,	Thursday,	" "
" "	1817,	Tuesday,	James Monroe.
" "	1821,	Sunday,	" "
" "	1825,	Friday,	John Q. Adams.
" "	1829,	Wednesday,	Andrew Jackson.
" "	1833,	Monday,	" "
" "	1837,	Saturday,	Martin Van Buren.
" "	1841,	Thursday,	Wm. H. Harrison.
" "	1845,	Tuesday,	James K. Polk.
" "	1849,	Sunday,	Zachary Taylor.
" "	1853,	Friday,	Franklin Pierce.
" "	1857,	Wednesday,	James Buchanan.
" "	1861,	Monday,	Abraham Lincoln.
" "	1865,	Saturday,	" "
" "	1869,	Thursday,	Ulysses S. Grant.
" "	1873,	Tuesday,	" "
" "	1877,	Sunday,	Rutherford B. Hays.
" "	1881,	Friday,	James A. Garfield.

March 4th,	1885,	Wednesday,	Grover Cleveland.
"	"	1889,	Monday,	Benjamin Harrison.
"	"	1893,	Saturday,	Grover Cleveland.
"	"	1897,	Thursday,	
"	"	1901,	Monday,	
"	"	1905,	Saturday,	
"	"	1909,	Thursday,	
"	"	1913,	Tuesday,	
"	"	1917,	Sunday,	
"	"	1921,	Friday,	
"	"	1925,	Wednesday,	
"	"	1929,	Monday,	
"	"	1933,	Saturday,	
"	"	1937,	Thursday,	
"	"	1941,	Tuesday,	
"	"	1945,	Sunday,	
"	"	1949,	Friday,	
"	"	1953,	Wednesday,	
"	"	1957,	Monday,	
"	"	1961,	Saturday,	
"	"	1965,	Thursday,	
"	"	1969,	Tuesday,	
"	"	1973,	Sunday,	
"	"	1977,	Friday,	
"	"	1981,	Wednesday,	
"	"	1985,	Monday,	
"	"	1989,	Saturday,	
"	"	1993,	Thursday.	

Any one understanding what has been said in a preceding chapter concerning the dominical letter, can very easily make out such a table without going through the process of making calculations for every year. As every succeeding year, or any day of the year, commences one day later in the week than the year preceding, and two days later in leap-year, which makes five days every four years, and as the Presidential term is four years, so every inaugural occurs five days later in the week than it did in the preceding term.

Now, as counting forward five days is equivalent to counting back two, it will be much more convenient to count back two days every term. There is one exception, however, to this rule ; the year which completes the century is reckoned as a common year (that is, three centuries out of four), consequently we count forward only four days or back three.

Commencing, then, with the second inaugural of Washington, which occurred on Monday, March 4, 1993, and counting back two days to Saturday in 1797, three days to Wednesday in 1801, and two days to 1805, and so on two days every term till 1901, when, for reasons already given, we count back three days again for one term only, after which it will be two days for the next two hundred years ; hence anyone can make his calculations as he writes, and as fast as he can write. See table on 61st page.

SOME PECULIARITIES CONCERNING EVENTS WHICH FALL ON THE TWENTY-NINTH DAY OF FEBRUARY.

The civil year and the day must be regarded as commencing at the same instant. We cannot well reckon
a fraction of a day, giving to February 28 days and 6
hours, making the following month to commence six
hours later every year; if so, then March, for example, in

    1888 would commence at 12 m. night.
    1889   "   "   " 6 a. m.
    1890   "   "   " 12 m.
    1891   "   "   " 6 p. m.
    1892   "   "   " 12 m. night,

again, and so on.

Instead of doing so, we wait until the fraction accu-
mulates to a whole day; then give to February 29 days,
and the year 366. Therefore, events which fall on the
29th of February cannot be celebrated annually, but
only quadrennially; and at the close of those centuries
in which the intercalations are suppressed only octen-
nially. For example: From the year 1696 to 1704,
1796 to 1804, and 1896 to 1904, there is no 29th day
of February; consequently no day of the month in the
civil year on which an event falling on the 29th of
February could be celebrated. Therefore, a person
born on the 29th of February, 1896, could celebrate
no birthday till 1904, a period of eight years.

In every common year February has 28 days, each
day of the week being contained in the number of days
in the month four times; but in leap-year, when February has 29 days, the day which begins and ends the

month is contained five times. Let us suppose that in a certain year, when February has 29 days, the month comes in on Friday; it also must necessarily end on Friday.

After four years it will commence on Wednesday, and end on Wednesday, and so on, going back two days in the week every four years, until after 28 years we come back to Friday again. This, as has already been explained, is the dominical or solar cycle. For example: February in

The year 4 has five Fridays.
  "   "   8 "   " Wednesdays.
  "   " 12 "   " Mondays.
  "   " 16 "   " Saturdays.
  "   " 20 "   " Thursdays.
  "   " 24 "   " Tuesdays.
  "   " 28 "   " Sundays.
  "   " 32 "   " Fridays.

So that after 28 years we come back to Friday again; and so on every 28 years, until change of style in 1582, when the Gregorian rule of intercalation being adopted by suppressing the intercalations in three centurial years out of four interrupts this order at the close of these three centuries. For example—1700, 1800 and 1900, after which the cycle of 28 years will be continued until 2100, and so on. The cycle being interrupted by the Gregorian rule of intercalation, causes all events which occur between 28 and 12 years of the close of the centuries to fall on the same day of the week again in 40 years; and those events that fall within 12 years of the close of these centuries, to fall on the same day of the week again in 12 years; after which the cycle of 28 years will be continued during the century. See following table :

1804	February has five		Wednesdays.
1808	"	"	" Mondays.
1812	"	"	" Saturdays.
1816	"	"	" Thursdays.
1820	"	"	" Tuesdays.
1824	"	"	" Sundays.
1828	"	"	" Fridays.
1832	"	"	" Wednesdays.
1836	"	"	" Mondays.
1840	"	"	" Saturdays.
1844	"	"	" Thursdays.
1848	"	"	" Tuesdays.
1852	"	"	" Sundays.
1856	"	"	" Fridays.
1860	"	"	" Wednesdays.
1864	"	"	" Mondays.
1868	"	"	" Saturdays.
1872	"	"	" Thursdays.
1876	"	"	" Tuesdays.
1880	"	"	" Sundays.
1884	"	"	" Fridays.
1888	"	"	" Wednesdays.
1892	"	"	" Mondays.
1896	"	"	" Saturdays.
1900			
1904	"	"	" Mondays.
1908	"	"	" Saturdays.
1912	"	"	" Thursdays.
1916	"	"	" Tuesdays.
1920	"	"	" Sundays.
1924	"	"	" Fridays.
1928	"	"	" Wednesdays.

It will be seen from this table that in 1804 February had five Wednesdays; and then again in 1832, 1860 and 1888; then suppressing the intercalation in the year 1900 suppresses the 29th of February; so opposite 1900 in the table is blank, and the 29th of February does not occur again till 1904, and the five Wednesdays do not occur again till 1928—that is, 40 years from 1888, when it last occurred.

Again taking the five Mondays which occurred first in this century, in 1808, and then again in 1836, 1864 and in 1892, you will see, for reasons already given, that it will occur again in 12 years, that is, in 1904; and so on with all the days of the week, when it will be seen what is peculiar concerning the 29th of February.

But attention is particularly called to the five Thursdays, which occur first in this century 1816, and then again in 1844 and 1872, the last date being within 28 years of the close of the century. Suppressing the intercalation suppresses the 29th of February; consequently the five Thursdays do not occur again till 1912, that is 40 years from the preceding date, after which the cycle will be continued for two hundred years.

Hence it may be seen that the dominical or solar cycle of 28 years is so interrupted at the close of these centuries by the suppression of the leap-year, that certain events do not occur again on the same day of the week under 40 years, while others are repeated again on the same day of the week in 12 years; also the number of years in the cycle, that is $28 + 12 = 40$.

And again the change of style in 1582, causes all events which occur between 28 and 8 years of that change, to fall again on the same day of the week in 36 years, and all that occur within 8 years of that change to be repeated again on the same day of the week in 8 years, after which the cycle of 28 years is continued for 100 years; also that the number of years in the cycle, that is, 28+8= 36.

CHAPTER VII.

RULES FOR FINDING THE DAY OF THE WEEK OF EVENTS THAT TRANSPIRED PRIOR TO THE CHRISTIAN ERA.

First, it should be understood that the year 4 is the first leap year in our era, reckoning from the year 1 B. C., which must necessarily be leap year; so that the odd numbers 1, 5, 9, 13, etc., are leap years. Hence every year that is divisible by four and one remainder, is leap year; if no remainder, it is the first year after leap-year; if 3, the second; if 2, the third, thus:

$$45 \div 4 = 11, \text{ remainder, } 1,$$
$$44 \div 4 \quad 11, \text{ no remainder,}$$
$$43 \div 4 \quad 10, \text{ remainder, } 3,$$
$$42 \div 4 \quad 10, \text{ remainder, } 2,$$
$$41 \div 4 = 10, \text{ remainder, } 1,$$

and so on, every year being divided by four and 1 remainder is leap-year of 366 days. It should be borne in mind that the same calendar was in use without any correction from the days of Julius Cæsar 46 B. C. to Pope Gregory XIII in 1582; consequently the method of finding the dominical letter is, in some respects, similar to the one already given on the 44th page. But in some respects the one is the reverse of the other, for we reckon backward and forward from a fixed point—the era; that is the numbers increase each way from the era. Also the dominical letters occur in the natural order of the letters in reckoning backward, but exactly the reverse in reckoning forward. See table on the 73d page, where the dominical letter is placed opposite each year from 45 B. C. to 45 A. D. Now we use the same number 3, because C, the third letter is dominical letter for the year 1 B. C., the point from which we reckon. But instead of taking the remainder, after dividing by 7, *from* 3 or 10, to find the number of the letter, as in Part Second, Chapter IV, (q. v.) we add the remainder *to* 3; hence we have the following :

RULE.

Divide the number of the given year by 4, neglecting the remainders, and add the quotient to the given number, divide this amount by 7, and add the remainder to 3, and that amount will give the number of the letter, calling A, 1 ; B, 2 ; C, 3, etc.; except the first year after leap-year, (which is the year exactly divisible by 4), the number of the letter is one less than is indicated by the rule.

This rule gives the dominical letter for January and February only, in leap-year, while the letter that precedes it, is the letter for the rest of the year. If the amount be greater than seven, we should reckon from A to A or B again.

It has already been stated in Part First, Chapter III, (q. v.), that a change was made by Augustus Cæsar about 8 B. C., in the number of days in the month; and, as this change effects the day of the week on which certain events fall, it becomes necessary that they should be presented as they were arranged by Julius Cæsar, and as corrected by Augustus. Julius Cæsar gave to February 29 days in common years, and in leap-year 30. This arrangement was the very best that could possibly be made, but, as has already been shown, it was interrupted to gratify the vanity of Augustus.

The left hand column in the table on the 72d page represents the number of days in each month from the days of Julius Cæsar to Augustus, a period of 37 years. The right hand column represents the number of days in the months as they now are, and have been since the change was made by Augustus, 8 B. C. Consequently the rule for finding the day of the week on which events have fallen for the 37 years prior to the last mentioned date, is not perfectly exact, and needs a little explanation here.

The rule itself is given, and fully explained in Part Second, Chapter V, (q. v.) but cannot be applied to the 37 years without some correction. In all the months marked with a star, events fall one day later in the week than that which is indicated by the rule. This should be borne in mind, and make the event one day later in the week than that which is found by the rule. For example, Julius Cæsar was assassinated on the 15th of March, 44 B. C. By giving to February 28 days the first day of March would fall on Wednesday, and, of course, the 15th would be Wednesday. But Cæsar gave to February 29 days, so that the first day of March fell on Thursday, and the 15th was Thursday.

Hence, every event from March to September will fall one day later in the week than the rule indicates. But the rule is applicable to September, for it will make no difference whether there are 29 days in February or 31 in August, there are the same number of days from February to September. But the 31 days in September will cause all events to fall one day later in the week during the month of October, but they co-incide again during the month of November. The order is interrupted again in December by giving 31 days to November. See following table:

As Arranged by Julius Caesar.	As Corrected by Augustus, & B. C.
January, 31	January, 31
February, 29	February, 28
March, 31*	March, 31
April, 30*	April, 30
May, 31*	May, 31
June, 30*	June, 30
July, 31*	July, 31
August, 30*	August, 31
September, 31	September, 30
October, 30*	October, 31
November, 31	November, 30
December, 30*	December, 31

Dominical Letter.	Year.	Dominical Letter.	Year.	Dominical Letter.	Year.	Dominical Letter.	Year.
	B. C.		B. C.		A. D.		A. D.
cb	45	b	22	b	1	c	23
a	44	ag	21	a	2	ba	24
g	43	f	20	g	3	g	25
f	42	e	19	fe	4	f	26
ed	41	d	18	d	5	e	27
c	40	cb	17	c	6	dc	28
b	39	a	16	b	7	b	29
a	38	g	15	ag	8	a	30
gf	37	f	14	f	9	g	31
e	36	ed	13	e	10	fe	32
d	35	c	12	d	11	d	33
c	34	b	11	cb	12	c	34
ba	33	a	10	a	13	b	35
g	32	gf	9	g	14	ag	36
f	31	e	8	f	15	f	37
e	30	d	7	ed	16	e	38
dc	29	c	6	c	17	d	39
b	28	ba	5	b	18	cb	40
a	27	g	4	a	19	a	41
g	26	f	3	gf	20	g	42
fe	25	e	2	e	21	f	43
d	24	dc	1	d	22	ed	44
c	23					c	45

PART THIRD.

CYCLES--JULIAN PERIOD--EASTER.

HEBREW CALENDAR.

CHAPTER I.

THE SOLAR CYCLE.

Cycle, (Latin *Cyclus*, ring or circle). The revolution of a certain period of time which finishes and re-commences perpetually. Cycles were invented for the purpose of chronology, and for marking the intervals in which two or more periods of unequal length are each completed a certain number of times, so that both begin exactly in the same circumstance as at first. Cycles used in chronology are three: The solar cycle, the lunar cycle, and the cycle of indiction.

The solar cycle is a period of time after which the same days of the year recur on the same days of the week. If every year contained 365 days, then every year would commence one day later in the week than the year preceding, and the cycle would be completed in seven years. For if the first day of January, in any given year, fall on Sunday, then the following year on Monday, the third on Tuesday, and so on to Sunday again in seven years.

But this order is interrupted in the Julian calendar every four years by giving to February 29 days, and conseqently the year 366. Now the number of years in the intercalary period being four and the days of the week being seven, their product is $4 \times 7 = 28$; twenty-eight years then is a period after which the first day of the year and the first day of every month recur again in the same order on the same day of the week. This period is called the solar cycle or the cycle of the sun, the origin of which is unknown; but is supposed to have been invented about the time of the Council of Nice, in the year of our Lord 325; but the first year of the cycle is placed by chronologists nine years before the commencement of the Christian era.

Hence the year of the cycle corresponding to any given year in the Julian calendar is found by the following rule: Add nine to the date and divide the sum by twenty-eight; the quotient is the number of cycles elapsed, and the remainder is the year of the cycle. Should there be no remainder, the proposed year is the twenty-eighth, or last of the cycle. Thus, for the year 1892, we have $(1892 + 9) \div 28 = 67$, remainder 25. Therefore, 67 is the number of cycles, and 25 the number in the cycle.

CHAPTER II.

THE LUNAR CYCLE.

The Lunar cycle, or the cycle of the moon, is a period of nineteen years, after which the new and full moons fall on the same days of the year as they did nineteen years before. This cycle was invented by Meton, a celebrated astronomer of Athens, and may be regarded as the masterpiece of ancient astronomy. In nineteen solar years there are 235 lunations, a number which, on being divided by nineteen, gives twelve lunations, with seven of a remainder, to be distributed among the years of the period.

The period of Meton, therefore, consisted of twelve years containing twelve months each, and seven years containing thirteen months each, and these last formed the third, fifth, eighth, eleventh, thirteenth, sixteenth, and nineteenth years of the cycle. As it had now been discovered that the exact length of the lunation is a little more than twenty-nine and a half days, it became necessary to abandon the alternate succession of full and deficient months; and, in order to preserve a more accurate correspondence between the civil month and the lunation, Meton divided the cycle into 125 full months of 30 days, and 110 deficient months of 29 days each. The number of days in the period was, therefore, 6940 ; for $(125 \times 30) + (110 \times 29) = 6940$.

In order to distribute the deficient months through the period in the most equable manner, the whole period may be regarded as consisting of 235 full months of thirty days, or 7050 days, from which 110 days are to be deducted; for (235×30)=7050; 7050−110=6940, as before. This gives one day to be suppressed in sixty-four, so that if we suppose the months to contain each thirty days, and then omit every sixty-fourth day in reckoning from the beginning of the period, those months in which the omission takes place will, of course, be the deficient months.

The number of days in the period being known, it is easy to ascertain its accuracy both in respect of the solar and lunar motions. The exact length of nineteen solar years is (365d, 5h, 48m, 49.62s.)×19=6939d, 14h, 27m, 42.78s.; hence, the period, which is exactly 6940 days, exceeds nineteen annual revolutions of the earth by a little more than nine and a half hours. On the other hand, the exact time of the synodic revolution of the moon is 29d, 12h, 44m, 2.87s.; 235 lunations, therefore, contain 235×(29d, 12h, 44m, 2.87s.)=(6939d, 16h, 31m, 14.45s.), so that the period exceeds 235 lunations by nearly seven and a half hours.

At the end of four cycles, or seventy-six years, the accumulation of the seven and a half hours of difference between the cycle and 235 lunations amounts to thirty hours, or one whole day and six hours. Calippus, therefore, in order to make a correction of the Metonic cycle, proposed to quadruple the period of Meton, and deduct one day at the end of that time by changing one of the full months into a deficient month. The period of Calippus, therefore, consisted of three Metonic cycles of 6940 days each, and a period of 6939 days; and its error in respect to the moon, consequently, amounted to only six hours, or to one day in 304 years. This period exceeds seventy-six true solar years by 14h, 9m, 8.88s., but coincides exactly with seventy-six Julian years; and in the time of Calippus the length of the solar year was almost universally supposed to be exactly 365¼ days.

CHAPTER III.

THE LUNAR CYCLE AND GOLDEN NUMBER.

In connecting the lunar month with the solar year, the framers of the ecclesiastical calendar adopted the period of Meton, or lunar cycle, which they supposed to be exact. A different arrangement has, however, been followed with respect to the distribution of the months. The lunations are supposed to consist of twenty-nine and thirty days alternately, or the lunar year of 354 days; and in order to make up nineteen solar years, six embolismic or intercalary months, of thirty days each, are introduced in the course of the cycle, and one of twenty-nine days is added at the end. This gives $(19 \times 354) + (6 \times 30) + 29 = 6935$ days, to be distributed among 235 lunar months.

But every leap-year one day must be added to the lunar month in which the 29th of February is included. Now if leap-year happened on the first, second or third year of the period, there will be five leap-years in the period, but only four when the first leap-year falls on the fourth. In the former case the number of days in the period becomes 6940, and in the latter 6939. The mean length of the cycle is, therefore, 6939¾ days, agreeing exactly with nineteen Julian years. By means of the lunar cycle the new moons of the calendar were indicated before the reformation in 1582. As the cycle restores these phenomena to the same days of the civil month, they will fall on the same days in any two years which occupy the same place in the cycle; consequently a table of the moon's phases for nineteen years will serve for any year whatever when we know its number in the cycle.

The number of the year in the cycle is called the Golden Number; either because it was so termed by the Greeks, who, on account of its utility, ordered it to be inscribed in letters of gold in their temples, or more probably because it was usual to distinguish it by red letters in the calendar. The Golden Numbers were introduced into the calendar about the year 530, but disposed as they would have been if they had been inserted at the time of the Council of Nice. The cycle is supposed to commence with the year in which the new moon falls on the first day of January, which took place the year preceding the commencement of our era.

Hence to find the Golden Number for any year, we have the following rule: Add one to the date, divide the sum by nineteen; the quotient is the number of cycles elapsed, and the remainder is the Golden Number. Should there be no remainder, the proposed year is, of course, the last or nineteenth of the cycle. Thus, for the year 1892, we have $(1892+1) \div 19 = 99$, remainder 12; therefore, 99 is the number of cycles, and 12 the number in the cycle, or the Golden number.

It ought to be remarked that the new moons determined in this manner, may differ from the astronomical new moons sometimes as much as two days. The reason is, that the sum of the solar and lunar inequalities which are compensated in the whole period, may amount in certain cases to 10 degrees and thereby cause the new moon to arrive on the second day before or after its mean time.

The cycle of the sun brings back the days of the month to the same day of the week ; the cycle of the moon restores the new moons to the same day of the month ; therefore, 28 × 19 = 532 years, includes all the variations in respect of the new moons and the dominical letter, and is consequently a period after which the new moons again occur on the same day of the month and the same day of the week. This is called the Dionysian or Great Paschal Period, from its having been employed by Dionysius Exiguus in determining Easter Sunday.

CHAPTER IV.

CYCLE OF INDICTION, AND THE JULIAN PERIOD.

The cycle of Indiction or Roman Indiction, is a period of fifteen years ; not astronomical like the two former, but entirely arbitrary. Its origin and the purpose for which it was established are alike uncertain ; but it is conjectured that it was introduced by Constantine the Great, about the year 312 of the common era, and had reference to certain judicial acts that took place under the Greek emperors at stated intervals of fifteen years. In chronological reckoning, it is considered as having commenced on the first day of January of the year 313.

By extending it backward, it will be found that the cycle commenced three years before the beginning of our era. In order, therefore, to find the number of any year in the cycle of indiction, we have this rule: Add three to the date, divide the number by fifteen; and the remainder is the year of the indiction. Should there be no remainder, the proposed year is the fifteenth or last of the cycle. Thus, for the year 1892, we have $(1892 + 3) \div 15 = 126$, remainder 5. Therefore, 5 is the number in the cycle.

The Julian period, proposed by the celebrated Joseph Scaliger, as an universal measure of chronology, is a period of 7980 years, and is formed by the continual multiplication of the three numbers, 28, 19 and 15; that is, of the cycle of the sun, the cycle of the moon, and the cycle of indiction. Thus, $(28 \times 19 \times 15) = 7980$. In the course of this long period no two years can be expressed by the same numbers in all the three cycles.

The first year of the Christian era had 10 for its number in the cycle of the sun, 2 in the cycle of the moon, and 4 in the cycle of the indiction. Now, it is found by actual calculation, that the only number less than 7980 which, on being divided successively by 28, 19, and 15, leaves the respective remainders 10, 2 and 4, is 4714. Hence, the first year of the Christian era corresponded with the year 4714 of the Julian period, which period coincides with the 710th before the common mundane era 4004 B. C.; for $4004 + 710 = 4714$. Hence, also, the year of our era corresponding to any other year of the period, or *vice versa*, is found by the following rule:

When the given year is anterior to the commencement of the era, subtract the number of the year of the Julian period from 4714, and the remainder is the year before Christ; or, subtract the year before Christ from 4714, and the remainder is the corresponding year in the Julian period. Thus, Rome was founded in the year 3960 of the Julian period. What was the year before Christ? We have then $4714—3960=754$. Julius Cæsar was assassinated 44 years before Christ, what was the corresponding year of the Julian period? We have then, $4714—44=4670$.

When the given year is after Christ, subtract 4713 from the year of the period, and the remainder is the year of the era; or add 4713 to the year of the era, and the sum is the corresponding year in the Julian period. Thus, the Council of Nice was convened in the year 5038 of the Julian period, what was the year of our era? We have then, $5038—4713=325$. What year of the Julian period corresponds with the present year, 1892? We have then, $4713+1892=6605$.

CHAPTER V.

EASTER.

Easter (Germ. *Ostern*, Old Saxon *Oster*, from *Osten*, signifying rising). The English name is probably derived from Ostera or Eostre, the Teutonic goddess of spring, whose festival occurred about the same time of the year as the celebration of Easter. The Hebrew-Greek word Pascha has passed into the name given to this feast by most Christian nations. This festival is held in commemoration of our Lord's resurrection.

The Jews celebrated their passover, in conformity with the directions given them by Moses, on the 14th day of the month Nisan, being the lunar month of which the 14th day either falls on or next follows the day of the vernal equinox. In the year of our Lord's crucifixion this fell on a Friday; the resurrection, therefore, took place on the first day of the week, which from thence is denominated the Lord's Day.

The primitive Christians, in celebrating this anniversary, fell into two different systems. The Western churches observed the nearest Sunday to the full moon of Nisan, taking no account of the day on which the passover would be celebrated. The Asiatics, on the other hand, following the Jewish calendar, adopted the 14th of Nisan upon which to commemorate the crucifixion, and observed the festival of Easter on the third day following, upon whatever day of the week that might fall, hence they obtained the name of Quartodecimans, (from quarto, four, and decem, ten,) the fourteenth day men. The former appealed to the authority of St. Peter and St. Paul, the latter to that of St John.

The dispute which took place upon this point in the second and third centuries of our era is remarkable, as connected with perhaps the first event which can be brought to bear upon the question of the primacy of the Roman bishop; and it is the more interesting as both parties are accustomed to claim it as a testimony in favor of their own views. Victor, bishop of Rome, wrote an imperious letter to the Asiatic bishops, requiring their conformity to the Western rule; which was answered by Polycrates, bishop of Ephesus, in the name of the rest, expressing their resolution to maintain the custom handed down to them by their ancestors. The Roman bishop thereupon broke off communion with them; but he was rebuked by Irenaeus, of Lyons, and it was agreed by his mediation that each party should retain its customs. Such continued to be the practice till the time of Constantine, when the Council of Nice determined the matter by the following Canons:

a—Easter must be celebrated on a Sunday.

b—This Sunday must follow the 14th day of the paschal moon, so that if the 14th day of the paschal moon fall on a Sunday, then Easter must be celebrated on the Sunday following.

c—The paschal moon is that moon of which the 14th day either falls on or next follows the day of the vernal equinox.

d—The 21st of March is to be accounted the day of the vernal equinox. (Appendix L.)

Sometimes a misunderstanding has arisen from not observing that this regulation is to be construed according to the tabular full moon as determined from the epact, and not by the true full moon, which in general, occurs one or two days earlier. From these conditions it follows, that if the paschal full moon fall on Saturday, the 21st of March, then Easter will happen on the 22d, its earliest possible date. For if the full moon arrive on the 20th, it would not be the paschal full moon, which cannot happen before the 21st, consequently the following moon is the paschal full moon, which happens 30 days after the 20th of March, which is the 19th of April. Now, if in this case the 19th of April is Sunday, then Easter must be celebrated the following Sunday, or the 26th of April. Hence, Easter Sunday cannot happen earlier than the 22d of March, or later than the 26th of April.

The observance of these rules renders it necessary to reconcile three periods which have no common measure, namely, the week, the lunar month, and the solar year; and as this can be done only approximately, and within certain limits, the determination of Easter is an affair of considerable nicety and complication. It has already been shown that the lunar cycle contained 6939 days and 18 hours; also, that the exact time of 235 lunations is 6939d, 16h, 31m, 14.45s. The difference, which is 1h, 28m, 45.55s., amounts to a day in 308 years, so that at the end of this time the new moons occur one day earlier than they are indicated by the Golden Numbers. During the 1257 years that elapsed between the Council of Nice and the reformation, the error had accumulated to four days, so that the new moons, which were marked in the calendar as happening, for example, on the 5th of the month, actually fell on the 1st.

It would have been easy to correct this error by placing the Golden Numbers four lines higher in the new calendar, but the suppression of ten days had already rendered it necessary to place them ten lines lower, and to carry those which belonged, for example, to the 5th and 6th of the month, to the 15th and 16th. But supposing this correction to have been made, it would have become necessary, at the end of 308 years, to place them one line higher, in consequence of the accumulation of the error of the cycle to a whole day. On the other hand, as the Golden Numbers were only adapted to the Julian calendar, every omission of the centenary intercalation would require them to be placed one line lower, opposite the 6th, for example, instead of the 5th of the month, so that, generally speaking, the places of the Golden Numbers would have to be changed every century. On this account Lilius thought fit to reject the Golden Numbers from the Calendar, and supply their places by another set of numbers called Epacts, the use of which we shall now proceed to explain.

Epact, (Greek *epaktos*, added or introduced). The excess of the solar year beyond the lunar, employed in the calendar to signify the moon's age at the beginning of the year. The common solar year consisted of 365 days and the lunar year only 354 days, the difference is eleven; whence, if a new moon fall on the first day of January in any year, the moon will be eleven days old on the first day of the following year, and

twenty-two days old on the first of the third year. The
numbers eleven and twenty-two are therefore the epacts
of those years respectively. Another addition of eleven
gives thirty-three for the epact of the fourth year; but
in consequence of the insertion of the intercalary month
in each third year of the lunar cycle, this epact is reduced to three; for $33-30=3$. In like manner the
epacts of all the following years of the cycle are obtained by successively adding eleven to the epact of the
former year, and rejecting thirty as often as the sum
exceeds or equals that number.

In order to show how the epacts are connected with
the Golden Numbers, let a cypher represent the new
moon on the first day of January in any year, then the
epacts and Golden Numbers for a whole lunar cycle
would be represented thus:

1	2	3	4	5	6	7	8	9
0	11	22	3	14	25	6	17	28

10	11	12	13	14	15	16	17	18	19
9	20	1	12	23	4	15	26	7	18

But the order is interrupted at the end of the cycle;
for the epact of the following year found in the same
manner would be $18+11=29$, whereas it ought to be
a cipher to correspond with the moon's age and the
Golden Number 1. The reason for this is, that the intercalary month, inserted at the end of the cycle, contains only twenty-nine days instead of thirty; whence,
after 11 has been added to the epact of the year corresponding to the Golden Number 19, we must reject
twenty-nine instead of thirty, in order to have the epact
of the succeeding year; or, which comes to the same
thing, we must add twelve to the epact of the last year
of the cycle, and then reject thirty as before. Thus,

18 + 12 = 30; 30 — 30 = 0; the cipher corresponding with the Golden Number 1.

This method of forming the epacts might have been continued indefinitely if the Julian intercalation had been followed without correction and the cycle had been perfectly exact; but as neither of these suppositions is true, two equations or corrections must be applied, one depending on the error of the Julian year, which is called the solar equation; the other on the errors of the lunar cycle, which is called the lunar equation. The solar equation occurs three times in 400 years, namely, in every secular year which is not a leap-year; for in this case the omission of the intercalary day causes the new moons to arrive one day later in all the following months, so that the moon's age at the end of the month is one day less than it would have been if the intercalation had been made, and the epacts must accordingly be all diminished by unity. Thus, the epacts 11, 22, 3, 14, etc., become 10, 21, 2, 13, etc.

On the other hand, when the time by which the new moons anticipate the lunar cycle amounts to a whole day, which, as we have seen, it does in 308 years, the new moons will arrive one day earlier and the epacts must, consequently, be increased by unity. Thus, the epacts 11, 22, 3, 14, etc., in consequence of the lunar equation, becomes 12, 23, 4, 15, etc. In order to preserve the uniformity of the calendar, the epacts are changed only at the commencement of the century; the correction of the error of the lunar cycle is therefore made at the end of 300 years. In the Gregorian calendar this error is assumed to amount to a day in

312½ years, or eight days in 2500 years, an assumption which requires the line of epacts to be changed seven times successively at the end of each period of 300 years, and once at the end of 400 years ; and from the manner in which the epacts were disposed at the reformation, it was found most correct to suppose one of the periods of 2500 years to terminate with the year 1800.

The years in which the solar equation occurs, counting from the reformation, are 1700, 1800, 1900, 2100, 2200, 2300, 2500, etc. Those in which the lunar equation occurs are 1800, 2100, 2400, 2700, 3000, 3300, 3600, 3900, after which 4300, 4600, and so on. When the solar equation occurs, the epacts are diminished by unity ; when the lunar equation occurs, the epacts are augmented by unity ; and when both equations occur together, as in 1800, 2100, 2700, etc., they compensate each other, and the epacts are not changed.

CHAPTER VI.

A NEW AND EASY METHOD OF FIXING THE DATE OF EASTER.

In determining the date of Easter, we make use of the numbers called epacts ; and, as these numbers have already been explained in the preceding chapter, (q. v.) it will be necessary to give them only a brief notice here. Epact, as has already been defined, is the excess of the solar year beyond the lunar, employed in the calendar to signify the moon's age at the beginning of the year; that is, if a new moon fall on the first day of

January in any year, it will be eleven days old on the first day of the following year, and twenty-two days old on the first day of the third year, and so on.

Now, in this work, in fixing the date of Easter, we abandon the use of the new moons altogether, and make calculations wholly from the paschal full moons, which cannot happen earlier than the 21st of March, nor later than the 19th of April. Appendix J. The epacts are here used to show the day of the month on which the paschal full moons fall ; that is, if the paschal moon fall on a given day of the month in any year, it will happen eleven days earlier the following year, and twenty-two days earlier the third year, and so on. To illustrate, suppose the paschal moon fall on the 18th of April in any given year, on the following year it would fall on the 7th, in the third year on the 27th of March ; and in the fourth year the moon would full on the 16th of March, but that would not be the paschal moon, which cannot happen earlier than the 21st ; so the following moon would be the paschal moon, which happens thirty days later, or the 15th of April ; then the fifth year it would fall on the 4th of April, and so on.

The solar and lunar equations or corrections are not made by change of epacts, for only one line of epacts is used in this work, but these corrections are made by a change of the day of the month on which the cycle commences. This change is made at the beginning of a century, and, of course, does not occur but once in a hundred years, and frequently no change is made for two, and even three hundred years. The reason for making these changes has been given in the preceding chapter, (q. v.), and will again be noticed in the proper place. The line of epacts used are thus represented, commencing with a cipher as the point of departure : 0, 11, 22, 3, 14, 25, 6, 17, 28, 9, 20, 1, 12, 23, 4, 15, 26, 7, 18. It should be borne in mind that the epacts are obtained by successively adding eleven to the epact of the former year, and rejecting thirty as often as the sum exceeds or equals that number. But, as the intercalary month inserted at the end of the cycle contains only 29 days, add twelve instead of eleven, to eighteen, the last of the cycle, and then reject thirty as before ; thus, $18 + 12 = 30$; then $30 - 30 = 0$. The cycle being completed, we again commence with the cipher as the point of departure.

After having found the paschal full moons for one lunar cycle, a period of nineteen years, then the paschal moons again occur in the same order, on the same days of the month, as they did nineteen years before. Now, as has also been shown in the preceding chapter, this cycle might have been continued indefinitely had the Julian intercalation been followed without correction, and the cycle been perfectly exact ; but neither of these being true, two equations or corrections must be made, one depending on the error of the Julian calendar, which is called the solar-equation ; the other on the error of the lunar cycle, which is called the lunar equation.

Every omission of the intercalary day, which occurs three times in 400 years, will cause the full moons to fall one day later ; for example, on the 13th of the month instead of the 12th. On the other hand, as has also been shown in the preceding chapter, the error of the lunar cycle is one day in 300 years ; so that at the end of every 300 years the full moons will fall one day earlier, for example, on the 11th of the month instead of the 12th. Now, when both equations occur together, they compensate each other ; that is, while the solar equation would cause the full moon to fall on the 13th, the lunar equation would make it fall on the 11th ; therefore, no correction is to be made—there is nothing to correct. Had they occurred singly, the full moon, at the beginning of the cycle, would have fallen either on the 13th or the 11th ; but as they occur together, no change is made; and the full moons of the calendar will remain as they are for the next one hundred years.

Hence, the date of Easter may very easily be determined, as indicated in the following tables (q. v.). It is known by actual calculation that the paschal full moon fell on the 12th of April in the year 1596, which moon was the first of a cycle after the reformation of the calendar by Gregory. Now, by taking the epact of the following years of the cycle, which are 11, 22, 3, 14, 25, etc., from 12, the date of the first paschal moon, and you will have all the moons of the cycle. Of course, the epacts 22, 25, etc., cannot be taken from 12, but being carried back from the 12th of April, they will show on what day in March the full moons fall. When the epacts are greater than 12, it would be more convenient to take them from 43, as the number of days in March being 31, so $12+31=43$.

To find the paschal moons of the cycle, we have then this rule : If the epact is less than 12, take it from 12 ; if greater, take it from 43, and the remainder will be the date of the paschal moon ; unless the full moon fall before the 21st of March, in which case the following moon will be the paschal moon, which happens thirty days later. But when the solar equation occurs in 1710, causing the cycle to commence with the 13th of April, then the epacts must be taken from 13, or $13+31=44$. And again in 1900, the correction makes the cycle commence on the 14th of April ; so the number from which the epacts are taken is 14, or $14+31=$ 45, and so on. Whenever there is a change of date of the paschal moon in the beginning of the cycle, as there is again in 2204, 2318 and 2413, etc., as may be seen in the following tables, then the epacts must be taken from that date, or that date plus 31, the number of days in March.

Or the date of the paschal moons may very easily be determined by taking eleven successively from the date of every preceding full moon, and that will give the date of the paschal moons ; only it should be borne in mind that, whenever the full moon falls before the 21st of March, the following moon is the paschal moon, which happens thirty days later.

As Easter Sunday is the first Sunday after the paschal full moon, so all that remains to be done in fixing the date of Easter, is to find the day of the month on which that Sunday falls ; and as this can easily be done by the use of the dominical letter, which letter and its use in fixing dates having been fully explained in Part Second, Chapters IV and V, (q. v.), a repetition seems to be unnecessary here.

Dominical Letter.	Year.	Paschal Full Moon.	Easter.	Epact.	Golden Number.	Dominical Letter.	Year	Paschal Full Moon.	Easter.	Epact.	Golden Number.
gf	1596	April 12	14	0	1	d	1615	April 12	19	0	1
e	1597	" 1	6	11	2	cb	1616	" 1	3	11	2
d	1598	March 21	22	22	3	a	1617	March 21	26	22	3
c	1599	April 9	11	3	4	g	1618	April 9	15	3	4
ba	1600	March 29	2	14	5	f	1619	March 29	31	14	5
g	1601	April 17	22	25	6	ed	1620	April 17	19	25	6
f	1602	" 6	7	6	7	c	1621	" 6	11	6	7
e	1603	March 26	30	17	8	b	1622	March 26	27	17	8
dc	1604	April 14	18	28	9	a	1623	April 14	16	28	9
b	1605	" 3	10	9	10	gf	1624	" 3	7	9	10
a	1606	March 23	26	20	11	e	1625	March 23	30	20	11
g	1607	April 11	15	1	12	d	1626	April 11	12	1	12
fe	1608	March 31	6	12	13	c	1627	March 31	4	12	13
d	1609	April 19	26	23	14	ba	1628	April 19	23	23	14
c	1610	" 8	11	4	15	g	1629	" 8	15	4	15
b	1611	March 28	3	15	16	f	1630	March 28	31	15	16
ag	1612	April 16	22	26	17	e	1631	April 16	20	26	17
f	1613	" 5	7	7	18	dc	1632	" 5	11	7	18
e	1614	March 25	30	18	19	b	1633	March 25	27	18	19

By close examination of the above tables, it will be seen that there is just eleven days difference in the date of these paschal moons, from year to year, through the whole lunar cycle, and through all lunar cycles. In determining the date of Easter, it will also be seen, that whenever the full moon falls before the 21st of March, then the following moon, which happens thirty days later, is the paschal moon, as the 21st of March is its earliest possible date. Also when the cycle is

Dominical Letter.	Year.	Paschal Full Moon.	Easter.	Epact.	Golden Number.	Dominical Letter.	Year.	Paschal Full Moon.	Easter.	Epact.	Golden Number.
a	1634	April 12	16	0	1	e	1653	April 12	13	0	1
g	1635	" 1	8	11	2	d	1654	" 1	5	11	2
fe	1636	March 21	23	22	3	c	1655	March 21	28	22	3
d	1637	April 9	12	3	4	ba	1656	April 9	16	3	4
c	1638	March 29	4	14	5	g	1657	March 29	1	14	5
b	1639	April 17	24	25	6	f	1658	April 17	21	25	6
ag	1640	" 6	8	6	7	e	1659	" 6	13	6	7
f	1641	March 26	31	17	8	dc	1660	March 26	28	17	8
e	1642	April 24	20	28	9	b	1661	April 14	17	28	9
d	1643	" 3	5	9	10	a	1662	" 3	9	9	10
cb	1644	March 23	27	20	11	g	1663	March 23	25	20	11
a	1645	April 11	16	1	12	fe	1664	April 11	13	1	12
g	1646	March 31	1	12	13	d	1665	March 31	5	12	13
f	1647	April 19	21	23	14	c	1666	April 19	25	23	14
ed	1648	" 8	12	4	15	b	1667	" 8	10	4	15
c	1649	March 28	4	15	16	ag	1668	March 28	1	15	16
b	1650	April 16	17	26	17	f	1669	April 16	21	26	17
a	1651	" 5	9	7	18	e	1670	" 5	6	7	18
gf	1652	March 25	31	18	19	d	1671	March 25	29	18	19

completed, then the paschal moons again occur in the same order, on the same day of the month as they did nineteen years before. Now this cycle is six times repeated in a period of 114 years, when the intercalary day being suppressed in 1700, causes the first paschal moon of the cycle to fall on the 13th of April instead of the 12th, and all the moons of the cycle to fall one day later than they would had the correction not been made. The cycle is now repeated ten times without

Dominical Letter	Year	Paschal Full Moon	Easter	Epact	Golden Number	Dominical Letter	Year	Paschal Full Moon	Easter	Epact	Golden Number
cb	1672	April 12	17	0	1	g	1691	April 12	15	0	1
a	1673	" 1	2	11	2	fe	1692	" 1	6	11	2
g	1674	March 21	25	22	3	d	1693	March 21	22	22	3
f	1675	April 9	14	3	4	c	1694	April 9	11	3	4
ed	1676	March 29	5	14	5	b	1695	March 29	3	14	5
c	1677	April 17	18	25	6	ag	1696	April 17	22	25	6
b	1678	" 6	10	6	7	f	1697	" 6	7	6	7
a	1679	March 26	2	17	8	e	1698	March 26	30	17	8
gf	1680	April 14	21	28	9	d	1699	April 14	19	28	9
e	1681	" 3	6	9	10	c	1700	" 3	4	9	10
d	1682	March 23	29	20	11	b	1701	March 23	27	20	11
c	1683	April 11	18	1	12	a	1702	April 11	16	1	12
ba	1684	March 31	2	12	13	g	1703	March 31	1	12	13
g	1685	April 19	22	23	14	fe	1704	April 19	20	23	14
f	1686	" 8	14	4	15	d	1705	" 8	12	4	15
e	1687	March 28	30	15	16	c	1706	March 28	4	15	16
dc	1688	April 16	18	26	17	b	1707	April 16	17	26	17
b	1689	" 5	10	7	18	ag	1708	" 5	8	7	18
a	1690	March 25	26	18	19	f	1709	March 25	31	18	19

correction, that is, till the year 1900, a period of 190 years, when the intercalation being again suppressed, causes the first paschal moon of the cycle to fall on the 14th of April, and, of course, all the other moons of the cycle to fall one day later. The reason the correction is not made the first year of the century is, the lunar cycle must first be completed, and that did not occur until 1710. As 100 is not a multiple of 19, the number of years in the cycle, and, as the corrections

Dominical Letter	Year	Paschal Full Moon	Easter	Epact	Golden Number	Dominical Letter	Year	Paschal Full Moon	Easter	Epact	Golden Number
e	1710	April 13	20	0	1	b	1729	April 13	17	0	1
d	1711	" 2	5	11	2	a	1730	" 2	9	11	2
cb	1712	March 22	27	22	3	g	1731	March 22	25	22	3
a	1713	April 10	16	3	4	fe	1732	April 10	13	3	4
g	1714	March 30	1	14	5	d	1733	March 30	5	14	5
f	1715	April 18	21	25	6	c	1734	April 18	25	25	6
ed	1716	" 7	12	6	7	b	1735	" 7	10	6	7
c	1717	March 27	28	17	8	ag	1736	March 27	1	17	8
b	1718	April 15	17	28	9	f	1736	April 15	21	28	9
a	1719	" 4	9	9	10	e	1738	" 4	6	9	10
gf	1720	March 24	31	20	11	d	1739	March 24	29	20	11
e	1721	April 12	13	1	12	cb	1740	April 12	17	1	12
d	1722	" 1	5	12	13	a	1741	" 1	2	12	13
c	1723	March 21	28	23	14	g	1742	March 21	25	23	14
ba	1724	April 9	16	4	15	f	1743	April 9	14	4	15
g	1725	March 29	1	15	16	ed	1744	March 29	5	15	16
f	1726	April 17	21	26	17	c	1745	April 17	18	26	17
e	1727	" 6	13	7	18	b	1746	" 6	10	7	18
dc	1728	March 26	28	18	19	a	1747	March 26	2	18	19

cannot be made only at the beginning of the cycle, so
they cannot be made the first year of the century only
once in 1900 years. It may be seen from one of the
above tables that the correction is made in the year
1900, for the reason that that is the first century which
is a multiple of 19. The next centurial year that is
exactly divisible by 19, is 3800. Therefore, none of
the corrections for the next 1900 years, will occur on
the first year of the century. It may also be seen from

Dominical Letter	Year	Paschal Full Moon	Easter	Epact	Golden Number	Dominical Letter	Year	Paschal Full Moon	Easter	Epact	Golden Number
gf	1748	April 13	14	0	1	d	1767	April 13	19	0	1
e	1749	" 2	6	11	2	cb	1768	" 2	3	11	2
d	1750	March 22	29	22	3	a	1769	March 22	26	22	3
c	1751	April 10	11	3	4	g	1770	April 10	15	3	4
ba	1752	March 30	2	14	5	f	1771	March 30	31	14	5
g	1753	April 18	22	25	6	ed	1772	April 18	19	25	6
f	1754	" 7	14	6	7	c	1773	" 7	11	6	7
e	1755	March 27	30	17	8	b	1774	March 27	3	17	8
dc	1756	April 15	18	28	9	a	1775	April 15	16	28	9
b	1757	" 4	10	9	10	gf	1776	" 4	7	9	10
a	1758	March 24	26	20	11	e	1777	March 24	30	20	11
g	1759	April 12	15	1	12	d	1778	April 12	19	1	12
fe	1760	" 1	6	12	13	c	1779	" 1	4	12	13
d	1761	March 21	22	23	14	ab	1780	March 21	26	23	14
c	1762	April 9	11	4	15	g	1781	April 9	15	4	15
b	1763	March 29	3	15	16	f	1782	March 29	31	15	16
ag	1764	April 17	22	26	17	e	1783	April 17	20	26	17
f	1765	" 6	7	7	18	dc	1784	" 6	11	7	18
e	1766	March 26	30	18	19	b	1785	March 26	27	18	19

the above tables, that, though the intercalary day was
suppressed in the year 1800, no change is made in the
date of the paschal moon. The reason is, the lunar
equation also occurred ; while the former correction
would cause the paschal moon to fall one day later,
that is on the 14th day of April, the latter would make
it fall one day earlier, that is on the 12th ; so they compensate each other, and there is no correction to be
made until the year 1900, when the solar equation

Dominical Letter	Year	Paschal Full Moon	Easter	Epact	Golden Number	Dominical Letter	Year	Paschal Full Moon	Easter	Epact	Golden Number
a	1786	April 13	16	0	1	f	1805	April 13	14	0	1
g	1787	" 2	8	11	2	e	1806	" 2	6	11	2
fe	1788	March 22	23	22	3	d	1807	March 22	29	22	3
d	1789	April 10	12	3	4	cb	1808	April 10	17	3	4
c	1790	March 30	4	14	5	a	1809	March 30	2	14	5
b	1791	April 18	24	25	6	g	1810	April 18	22	25	6
ag	1792	" 7	8	6	7	f	1811	" 7	14	6	7
f	1793	March 27	31	17	8	ed	1812	March 27	29	17	8
e	1794	April 15	20	28	9	c	1813	April 15	18	28	9
d	1795	" 4	5	9	10	b	1814	" 4	10	9	10
cb	1796	March 24	27	20	11	a	1815	March 24	26	20	11
a	1797	April 12	16	1	12	gf	1816	April 12	14	1	12
g	1798	" 1	8	12	13	e	1817	" 1	6	12	13
f	1799	March 21	24	23	14	d	1818	March 21	22	23	14
e	1800	April 9	13	4	15	c	1819	April 9	11	4	15
d	1801	March 29	5	15	16	ba	1820	March 29	2	15	16
c	1802	April 17	18	26	17	g	1821	April 17	22	26	17
b	1803	" 6	10	7	18	f	1822	" 6	7	7	18
ag	1804	March 26	1	18	19	e	1823	March 26	30	18	19

again occurs, and the first paschal moon of the cycle falls on the 14th ; which cycle is repeated sixteen times in a period of 304 years, after which, in 2204, the date of the first paschal moon is the 15th of April. The reason there is no correction to make in this long period is, first, the year 2000 is a common year in the Gregorian calendar ; second, in the year 2100 both the solar and the lunar equations again occur, and occurring together, they compensate each other ; conse-

Dominical Letter	Year	Paschal Full Moon	Easter	Epact	Golden Number	Dominical Letter	Year	Paschal Full Moon	Easter	Epact	Golden Number
dc	1824	April 13	18	0	1	a	1843	April 13	16	0	1
b	1825	" 2	3	11	2	gf	1844	" 2	7	11	2
a	1826	March 22	26	22	3	e	1845	March 22	23	22	3
g	1827	April 10	15	3	4	d	1846	April 10	12	3	4
fe	1828	March 30	6	14	5	c	1847	March 30	4	14	5
d	1829	April 18	19	25	6	ba	1848	April 18	23	25	6
c	1830	" 7	11	6	7	g	1849	" 7	8	6	7
b	1831	March 27	3	17	8	f	1850	March 27	31	17	8
ag	1832	April 15	22	28	9	e	1851	April 15	20	28	9
f	1833	" 4	7	9	10	dc	1852	" 4	11	9	10
e	1834	March 24	30	20	11	b	1853	March 24	27	20	11
d	1835	April 12	19	1	12	a	1854	April 12	16	1	12
cb	1836	" 1	3	12	13	g	1855	" 1	8	12	13
a	1837	March 21	26	23	14	fe	1856	March 21	23	23	14
g	1838	April 9	15	4	15	d	1857	April 9	12	4	15
f	1839	March 29	31	15	16	c	1858	March 29	4	15	16
ed	1840	April 17	19	26	17	b	1859	April 17	24	26	17
c	1841	" 6	11	7	18	ag	1860	" 6	8	7	18
b	1842	March 26	27	18	19	f	1861	March 26	31	18	19

quently the cycle is continued until 2204, after which, as has already been stated, the date of the first paschal moon is the 15th of April. This cycle is repeated six times in a period of 114 years, when in 2318, for reasons already given, the date of the first paschal moon of the next cycle falls on the 16th, and is repeated five times in a period of 95 years, when, in 2413, the lunar equation occurs alone, and the date of the first paschal moon for the next 95 years, that is till the year 2508,

Dominical Letter	Year	Paschal Full Moon	Easter	Epact	Golden Number	Dominical Letter	Year	Paschal Full Moon	Easter	Epact	Golden Number
e	1862	April 13	20	0	1	b	1881	April 13	17	0	1
d	1863	" 2	5	11	2	a	1882	" 2	9	11	2
cb	1864	March 22	27	22	3	g	1883	March 22	25	22	3
a	1865	April 10	16	3	4	fe	1884	April 10	13	3	4
g	1866	March 30	1	14	5	d	1885	March 30	5	14	5
f	1867	April 18	21	25	6	c	1886	April 18	25	25	6
ed	1868	" 7	12	6	7	b	1887	" 7	10	6	7
c	1869	March 27	28	17	8	ag	1888	March 27	1	17	8
b	1870	April 15	17	28	9	f	1889	April 15	21	28	9
a	1871	" 4	9	9	10	e	1890	" 4	6	9	10
gf	1872	March 24	31	20	11	d	1891	March 24	29	20	11
e	1873	April 12	13	1	12	cb	1892	April 12	17	1	12
d	1874	" 1	5	12	13	a	1893	" 1	2	12	13
c	1875	March 21	28	23	14	g	1894	March 21	25	23	14
ba	1876	April 9	16	4	15	f	1895	April 9	14	4	15
g	1877	March 29	1	15	16	ed	1896	March 29	5	15	16
f	1878	April 17	21	26	17	c	1897	April 17	18	26	17
e	1879	" 6	13	7	18	b	1898	" 6	10	7	18
dc	1880	March 26	28	18	19	a	1899	March 26	2	18	19

falls back to the 15th of April. After which the 16th, on account of the solar equation, is again the date of the first paschal moon of the cycle for another period of 95 years; that is till the year 2603, when the solar equation causes the first paschal moon to fall on the 17th, which cycle is repeated sixteen times during another period of 304 years, after which, in 2907, the correction makes the date of the first paschal moon of the

Dominical Letter.	Year.	Paschal Full Moon.	Easter.	Epact.	Golden Number.	Dominical Letter.	Year.	Paschal Full Moon.	Easter.	Epact.	Golden Number.
g	1900	April 14	15	0	1	e	1919	April 14	20	0	1
f	1901	" 3	7	11	2	dc	1920	" 3	4	11	2
e	1902	March 23	30	22	3	b	1921	March 23	27	22	3
d	1903	April 11	12	3	4	a	1922	April 11	16	3	4
cb	1904	March 31	3	14	5	g	1923	March 31	1	14	5
a	1905	April 19	23	25	6	fe	1924	April 19	20	25	6
g	1906	" 8	15	6	7	d	1925	" 8	12	6	7
f	1907	March 28	31	17	8	c	1926	March 28	4	17	8
ed	1908	April 16	19	28	9	b	1927	April 16	17	28	9
c	1909	" 5	11	9	10	ag	1928	" 5	8	9	10
b	1910	March 25	27	20	11	f	1929	March 25	31	20	11
a	1911	April 13	16	1	12	e	1930	April 13	20	1	12
gf	1912	" 2	7	12	13	d	1931	" 2	5	12	13
e	1913	March 22	23	23	14	cb	1932	March 22	27	23	14
d	1914	April 10	12	4	15	a	1933	April 10	16	4	15
c	1915	March 30	4	15	16	g	1934	March 30	1	15	16
ba	1916	April 18	23	26	17	f	1935	April 18	21	26	17
g	1917	" 7	8	7	18	ed	1936	" 7	12	7	18
f	1918	March 27	31	18	19	c	1937	March 27	28	18	19

cycle, the 18th of April, which cycle is continued without correction till the year 3116, a period of 209 years. By reference to the above tables, it will be seen that the solar and lunar equations occur together in the year 2700 and compensate each other ; also, that the year 2800 is a common year in the Gregorian calendar; consequently there is no correction to make from 2603 to 2907 ; also the two equations occur together

Dominical Letter	Year	Paschal Full Moon	Easter	Epact	Golden Number	Dominical Letter	Year	Paschal Full Moon	Easter	Epact	Golden Number
b	1938	April 14	17	0	1	f	1957	April 14	21	0	1
a	1939	" 3	9	11	2	e	1958	" 3	6	11	2
gf	1940	March 23	24	22	3	d	1959	March 23	29	22	3
e	1941	April 11	13	3	4	cb	1960	April 11	17	3	4
d	1942	March 31	5	14	5	a	1961	March 31	2	14	5
c	1943	April 19	25	25	6	g	1962	April 19	22	25	6
ba	1944	" 8	9	6	7	f	1963	" 8	14	6	7
g	1945	March 28	1	17	8	ed	1964	March 28	29	17	8
f	1946	April 16	21	28	9	c	1965	April 16	18	28	9
e	1947	" 5	6	9	10	b	1966	" 5	10	9	10
dc	1948	March 25	28	20	11	a	1967	March 25	26	20	11
b	1949	April 13	17	1	12	gf	1968	April 13	14	1	12
a	1950	" 2	9	12	13	e	1969	" 2	6	12	13
g	1951	March 22	25	23	14	d	1970	March 22	29	23	14
fe	1952	April 10	13	4	15	c	1971	April 10	11	4	15
d	1953	March 30	5	15	16	ba	1972	March 30	2	15	16
c	1954	April 18	25	26	17	g	1973	April 18	22	26	17
b	1955	" 7	10	7	18	f	1974	" 7	14	7	18
ag	1956	March 27	1	18	19	e	1975	March 27	30	18	19

again in the year 3000 and compensate each other, is
the reason there is no correction to make from 2907 to
3116, after which the first paschal moon falls on the
19th, and is repeated fifteen times in a period of 285
years, that is till the year 3401, when the correction
makes the 20th of April, the date of the full moon,
but that cannot be the paschal moon, which cannot happen later than the 19th; consequently the

Dominical Letter	Year	Paschal Full Moon	Easter	Epact	Golden Number	Dominical Letter	Year	Paschal Full Moon	Easter	Epact	Golden Number
dc	1976	April 14	18	0	1	a	1995	April 14	16	0	1
b	1977	" 3	10	11	2	gf	1996	" 3	7	11	2
a	1978	March 23	26	22	3	e	1997	March 23	30	22	3
g	1979	April 11	15	3	4	d	1998	April 11	12	3	4
fe	1980	March 31	6	14	5	c	1999	March 31	4	14	5
d	1981	April 19	26	25	6	ba	2000	April 19	23	25	6
c	1982	" 8	11	6	7	g	2001	" 8	15	6	7
b	1983	March 28	3	17	8	f	2002	March 28	31	17	8
ag	1984	April 16	22	28	9	e	2003	April 16	20	28	9
f	1985	" 5	7	9	10	dc	2004	" 5	11	9	10
e	1986	March 25	30	20	11	b	2005	March 25	27	20	11
d	1987	April 13	19	1	12	a	2006	April 13	16	1	12
cb	1988	" 2	3	12	13	g	2007	" 2	8	12	13
a	1989	March 22	26	23	14	fe	2008	March 22	23	23	14
g	1990	April 10	15	4	15	d	2009	April 10	12	4	15
f	1991	March 30	31	15	16	c	2010	March 30	4	15	16
ed	1992	April 18	19	26	17	b	2011	April 18	24	26	17
c	1993	" 7	11	7	18	ag	2012	" 7	8	7	18
b	1994	March 27	3	18	19	f	2013	March 27	31	18	19

moon that precedes it by thirty days, and which
falls on the 21st of March, is the date of the first paschal moon of the cycle which commences with the
year 3401. The day of the month on which Easter
Sunday has fallen or will fall, from the year 1596 to
2013, is already determined, and may be seen by reference to the above tables. From 2013 to 3401, the date
of Easter is determined for one lunar cycle only, at the

Dominical Letter	Year	Paschal Full Moon	Easter	Epact	Golden Number	Dominical Letter	Year	Paschal Full Moon	Easter	Epact	Golden Number
ag	2204	April 15	22	0	1	f	2318	April 16	21	0	1
f	2205	" 4	7	11	2	e	2319	" 5	6	11	2
e	2206	March 24	30	22	3	dc	2320	March 25	28	22	3
d	2207	April 12	19	3	4	b	2321	April 13	17	3	4
cb	2208	" 1	3	14	5	a	2322	" 2	9	14	5
a	2209	March 21	26	25	6	g	2323	March 22	25	25	6
g	2210	April 9	15	6	7	fe	2324	April 10	13	6	7
f	2211	March 29	31	17	8	d	2325	March 30	5	17	8
ed	2212	April 17	19	28	9	c	2326	April 18	25	28	9
c	2213	" 6	11	9	10	b	2327	" 7	10	9	10
b	2214	March 26	27	20	11	ag	2328	March 27	1	20	11
a	2215	April 14	16	1	12	f	2329	April 15	21	1	12
gf	2216	" 3	7	12	13	e	2330	" 4	6	12	13
e	2217	March 23	30	23	14	d	2331	March 24	29	23	14
d	2218	April 11	12	4	15	cb	2332	April 12	17	4	15
c	2219	March 31	4	15	16	a	2333	" 1	2	15	16
ba	2220	April 19	23	26	17	g	2334	March 21	25	26	17
g	2221	" 8	15	7	18	f	2335	April 9	14	7	18
f	2222	March 28	31	18	19	ed	2336	March 29	5	18	19

beginning of each period; for the reason that it was deemed unnecessary, because the paschal moons, the epacts, and the Golden Numbers are the same for every cycle in the period. Therefore, all that remains to be done is to find the day of the month on which the first Sunday, after the paschal moon, falls. The dominical letters for any period may very easily be found by counting backwards one letter each year for every com-

Dominical Letter	Year	Paschal Full Moon	Easter	Epact	Golden Number	Dominical Letter	Year	Paschal Full Moon	Easter	Epact	Golden Number
f	2413	April 15	21	0	1	ag	2508	April 16	22	0	1
e	2414	" 4	6	11	2	f	2509	" 5	7	11	2
d	2415	March 24	29	22	3	e	2510	March 25	30	22	3
cb	2416	April 12	17	3	4	d	2511	April 13	19	3	4
a	2417	" 1	2	14	5	cb	2512	" 2	3	14	5
g	2418	March 21	25	25	6	a	2513	March 22	26	25	6
f	2419	April 9	14	6	7	g	2514	April 10	15	6	7
ed	2420	March 29	5	17	8	f	2515	March 30	31	17	8
c	2421	April 17	18	28	9	ed	2516	April 18	19	28	9
b	2422	" 6	10	9	10	c	2617	" 7	11	9	10
a	2423	March 26	2	20	11	b	2518	March 27	3	20	11
gf	2424	April 14	21	1	12	a	2519	April 15	16	1	12
e	2425	" 3	6	12	13	gf	2520	" 4	7	12	13
d	2426	March 23	29	23	14	e	2521	March 24	30	23	14
c	2427	April 11	18	4	15	d	2522	April 12	19	4	15
ba	2428	March 31	2	15	16	c	2523	" 1	4	15	16
g	2429	April 19	22	26	17	ba	2524	March 21	26	26	17
f	2430	" 8	14	7	18	g	2525	April 9	15	7	18
e	2431	March 28	30	18	19	f	2526	March 29	31	18	19

mon year, and two for leap-year, the fourth letter being dominical letter for January and February and the fifth for the rest of the year ; thus, if G be dominical letter for any given year, we would have then, G, F, E, DC ; B, A, G, FE ; D, C, B, AG ; F, E, D, CB, etc. By counting these letters backwards, or in the tables, from the bottom of the column upwards, they will occur in alphabetical order. Again it may be seen by refer-

Dominical Letter.	Year.	Paschal Full Moon.	Easter.	Epact.	Golden Number.	Dominical Letter.	Year.	Paschal Full Moon.	Easter.	Epact.	Golden Number.
b	2603	April 17	24	0	1	b	2907	April 18	24	0	1
ag	2604	" 6	8	11	2	ag	2908	" 7	8	11	2
f	2605	March 26	31	22	3	f	2909	March 27	31	22	3
e	2606	April 14	20	3	4	e	2910	April 15	20	3	4
d	2607	" 3	5	14	5	d	2911	" 4	5	14	5
cb	2608	March 23	27	25	6	cb	2912	March 24	27	25	6
a	2609	April 11	16	6	7	a	2913	April 12	16	6	7
g	2610	March 31	1	17	8	g	2914	" 1	8	17	8
f	2611	April 19	21	28	9	f	2915	March 21	24	28	9
ed	2612	" 8	12	9	10	ed	2916	April 9	12	9	10
c	2613	March 28	4	20	11	c	2917	March 29	4	20	11
b	2614	April 16	17	1	12	b	2918	April 17	24	1	12
a	2615	" 5	9	12	13	a	2919	" 6	9	12	13
gf	2616	March 25	31	23	14	gf	2920	March 26	31	23	14
e	2617	April 13	20	4	15	e	2921	April 14	20	4	15
d	2618	" 2	5	15	16	d	2922	" 3	5	15	16
c	2619	March 22	28	26	17	c	2923	March 23	28	26	17
ba	2620	April 10	16	7	18	ba	2924	April 11	16	7	18
g	2621	March 30	1	18	19	g	2925	March 31	1	18	19

ence to these tables, that Easter occurs less frequently on the 22d of March, its earliest possible date, and the 25th of April, which has hitherto been considered its latest possible date, than any of the days intervening. It cannot happen on the 22d, only when the paschal moon falls on the 21st, and that day must be Saturday. It fell on the 22d, first, after the reformation of the calendar by Gregory in 1598 ; again in 1693, 1761,

Dominical Letter.	Year.	Paschal Full Moon.	Easter.	Epact.	Golden Number.	Dominical Letter.	Year.	Paschal Full Moon.	Easter.	Epact.	Golden Number.
ba	3116	April 19	23	0	1	d	3401	March 21	22	0	1
g	3117	" 8	15	11	2	c	3402	April 9	11	11	2
f	3118	March 28	31	22	3	b	3403	March 29	3	22	3
e	3119	April 16	20	3	4	ag	3404	April 17	22	3	4
dc	3120	" 5	11	14	5	f	3405	" 6	7	14	5
b	3121	March 25	27	25	6	e	3406	March 26	30	25	6
a	3122	April 13	16	6	7	d	3407	April 14	19	6	7
g	3123	" 2	8	17	8	cb	3408	" 3	10	17	8
fe	3124	March 22	23	28	9	a	3409	March 23	26	28	9
d	3125	April 10	12	9	10	g	3410	April 11	15	9	10
c	3126	March 30	4	20	11	f	3411	March 31	7	20	11
b	3127	April 18	24	1	12	ed	3412	April 19	26	1	12
ag	3128	" 7	8	12	13	c	3413	" 8	11	12	13
f	3129	March 27	31	23	14	b	3414	March 28	3	23	14
e	3130	April 15	20	4	15	a	3415	April 16	23	4	15
d	3131	" 4	5	15	16	gf	3416	" 5	7	15	16
cb	3132	March 24	27	26	17	e	3417	March 25	30	26	17
a	3133	April 12	16	7	18	d	3418	April 13	19	7	18
g	3134	" 1	8	18	19	c	3419	" 2	4	18	19

and in 1818. It has not occurred since, nor will not again till 2285, a period of 467 years. The reason that it does not occur on the 22d of March in this long period is, the paschal moon does not fall on the 21st, from the year 1900 to 2204, a period of 304 years. We refer to the tabular moon, not to the true or astronomical moon, which may occur on the 21st more than once in this long period.

CHAPTER VII.

CHURCH FEASTS AND FASTS WHOSE DATE DEPEND ON THE DATE OF EASTER.

Feasts, or Festivals, are days set apart by the church, either for the grateful memorial of the most remarkable events connected with the plan of redemption, or upon which to commemorate the actions and sufferings of such persons as have been most instrumental in carrying forward the designs of God for the salvation of mankind.

The ecclesiastical year is divided into eight seasons, namely : Advent-tide, Christmas-tide, Epiphany-tide, Lenten-tide, Easter-tide, Ascension-tide, Whitsun-tide, and Trinity-tide. The first day of each of these seasons has been, and is now observed by the church in commemoration of some of the most remarkable events connected with the plan of redemption. All these will be noticed in the order in which they occur in the ecclesiastical year, while many other days intervening, which are observed as feasts or fasts, will be given a passing notice.

a—Advent Sunday, which is the day nearest St. Andrew's Day (Nov. 30), or the first Sunday after the 26th of November, has been recognized since the sixth century as the first day of the ecclesiastical year.

Advent (Latin *Adventus*, the coming,) signifies the coming of our Saviour, the period of the approach of the nativity. As Advent-tide lasts from Advent Sunday to Christmas, the length of the season depends upon

the day of the month on which Advent Sunday falls.
As it may happen as early as the 27th of November or
as late as the 3d of December, so Advent-tide will contain no more than twenty-eight days nor less than
twenty-two. It should be borne in mind that, though
this festival is classed among the movable feasts, it does
not depend upon the date of Easter, but upon the day
of the month on which Advent Sunday falls. The four
Sundays before Christmas were made preparation days
for the festival of Christmas, and were called the first,
second, third, and fourth Sundays in Advent.

Ember days and Ember weeks are the four seasons
set apart by the Western church for special prayer and
fasting, and the ordination of clergy; known in the
church as *quatuor tempora*, (the four seasons.) The
Ember weeks are the weeks next following St. Lucy's
Day (Dec. 13th), the first Sunday in Lent, Whitsun
Day, and Holy Cross Day (Sept. 14th). The Wednesdays, Fridays and Saturdays of these weeks are the
Ember days distinctively. The name by some is supposed to be derived from a German word signifying
Abstinence; by others it is supposed to signify Ashes.

b—Christmas (from Christ and the Saxon *macss*,
signifying the mass and a feast), is a festival held in
commemoration of the nativity of our Saviour throughout nearly the whole of Christendom. It is occupied,
therefore, with the event (the incarnation) which forms
the center and turning point of the history of the world.
Though the day of Christ's birth cannot be ascertained
from the New Testament, nor from any other source,
yet the whole Christian world for more than 1300 years
have concurred in celebrating the nativity on the 25th

of December. This is the first of the four great feasts in the ecclesiastical year ; the other three are Easter, Ascension, and Whitsun Day. The length of Christmas-tide or season is twelve days, lasting from the 25th of December to Epiphany.

c—Epiphany(Greek *Epiphania, Theophania* or *Christophania,*) is a festival in commemoration of the manifestation of Jesus Christ to the world as the Son of God, and referring to the appearance of the star which announced our Saviour's birth to the Gentiles, and the visit of the Magi, or wise Men of the East to the infant Jesus. This festival is held on the 6th of January invariably, consequently is not a movable feast, though the length of Epiphany-tide depends upon the date of Easter. As Easter may happen as early as the 22d of March or as late as the 26th of April (a variation of thirty-five days), so Epiphany-tide may consist of no less than twelve days nor more than forty-seven, as the season always ends the day before Septuagesima Sunday. (See tables at the close of this chapter.)

Septuagesima, Sexagesima, Quinquagesima. There being exactly fifty days between the Sunday next before Lent and Easter Day, inclusive, that Sunday was termed Quinquagesima, i. e., the fiftieth ; and the two immediately preceding Sundays were called from the next round numbers Sexagesima, the sixtieth ; and Septuagesima, the seventieth.

The Paschal Season extends from Septuagesima Sunday to Low Sunday, a period of seventy days. It takes its name from the Paschal festival or Easter, whose services end with Low Sunday, the octave, or eighth day, of Easter. It begins with Septuagesima Sunday

because the church services then begin to prepare the
minds of the faithful for the services of Lent, which are
themselves the preparation for Easter. May not Septuagesima Sunday be so called because there are just
seventy days in the Paschal Season ?

Shrove tide literally means confession-time, and is
the name given to the days immediately preceding Ash
Wednesday. These days were so called because on
them, and especially on the last of them (Shrove Tuesday) people were accustomed to confess their sins as a
preparation for Lent. In most Roman Catholic countries
it begins with Quinquagesima Sunday, the Sunday before Lent.

Ash Wednesday, the first day of Lent, (Latin *dies
cinerum*, the day of Ashes), was so called because it
was customary on that day for penitents to appear in
sackcloth, upon which occasion ashes were sprinkled
upon them.

d—Lent, (Anglo-Saxon *lengten*. Perhaps from *lenc-
gan*, to lengthen, because at this season the days
lengthen) the forty days fast, is the preparation for
Easter, and is observed in commemoration of our
Lord's fast in the wilderness. In most languages the
name given to this fast signifies the number of days—
forty ; but our word Lent signifies the Spring Fast, for
Lenten-tide in the Anglo-Saxon language was the Season of Spring, in German, Lenz.

The six Sundays in the Lenten-tide of forty-six days
are not counted in the fast, as all Sundays in the year
are reckoned as feast days, because our Saviour arose
from the dead on the first day of the week.

Quadragesima Sunday, the first Sunday in Lent, is

so called by analogy with the three Sundays which precede Lent, and which (as has already been stated) are called respectively Septuagesima, seventieth ; Sexagesima, sixtieth ; Quinquagesima, fiftieth ; and then Quadragesima, fortieth ; in round numbers forty days before Easter.

Holy Week, the last week in Lent, called also Passion Week, because within it is commemorated our Lord's sufferings. The days specially solemnized are Palm Sunday, Spy Wednesday, Holy, or Maundy Thursday, and Good Friday.

Palm Sunday (Latin *Dominica Palmarium*, or *Dominica* in *Palmis*) is the name usually given the Sunday before Easter ; a day celebrated in commemoration of Christ's triumphal entry into Jerusalem, so called because the people who had come to the feast, when they heard that Jesus was coming, took branches of palm trees and went forth to meet him, and cried, " Hosanna ; blessed is the King of Israel that cometh in the name of the Lord."

Spy Wednesday, so called in allusion to the betrayal of Christ by Judas, or the day on which he made the bargain to deliver him into the hands of his enemies for thirty pieces of silver.

Maundy Thursday (from *Dies mandati*, mandate Thursday), so called either from the command given his disciples to love one another, or to commemorate the sacrament of His supper.

Good Friday, so called in acknowledgment of the benefit derived from the death of Christ.

The closing scenes in the life of Christ, the events of Wednesday, Thursday and Friday of Holy Week, are

events of much more importance than were ever before
crowded into any one week in the history of the world.
The betrayal on Wednesday, the institution of the
sacrament on Thursday night, also the words of our
Saviour as recorded in John's gospel, from the 14th to
the 17th chapters inclusive, the agony and the bloody
sweat in the garden, the arrest and trial during Thursday night and Friday morning, the crucifixion at the
third hour, the darkness that was over all the land
from the sixth to the ninth hour, and the last words of
the blessed Jesus, " It is finished," (tasted death for
every man) ; these we say, are events of more importance to man than were ever before crowded into any
one week in the world's history.

The prophets who prophesied of these things, in-
quired and searched diligently, "searching what, or what
manner of time the Spirit of Christ which was in them
did signify, when it testified beforehand the sufferings
of Christ and the glory that should follow." And
about an hour before this prophecy began to be fulfilled our Saviour uttered these words : " Verily,
verily, I say unto you, that ye shall weep and lament,
but the world shall rejoice ; and ye shall be sorrowful,
but your sorrow shall be turned into joy." It was
probably not more than an hour from the time these
words fell from the Saviour's lips, that he was arrested
and led away to undergo a trial ; cruel mocking and
scourging, crucifixion and death upon the cross.

Then cometh Joseph of Arimathea, bringing fine
linen, and Nicodemus with his hundred pounds of
myrrh and aloes, and they two took the body of the
Lord Jesus and wrapped it in the linen with the spices,

and laid it in Joseph's own new tomb, which he had
hewn out in a rock, wherein never man before was laid,
and rolled a great stone against the door of the sepulcher
and departed.   Thus endeth Passion Week.   While
the body of Jesus is in the sepulcher the world is rejoicing, and the disciples are weeping and lamenting, according to the words of the Saviour, "Ye shall weep
and lament, but the world shall rejoice."

> He dies! the friend of sinners dies!
> Lo! Salem's daughters weep around;
> A solemn darkness veils the skies,
> A sudden trembling shakes the ground.

e—Easter (German, *Ostern*, Old Saxon *Oster*, from
osten, signifying rising,) is a church festival in commemoration of the resurrection of our Lord from the dead.
But the apparent victory which the enemies of Christ
had gained was of short duration, the rejoicing of the
world did not long continue, the remaining words of
our Saviour must needs be fulfilled :   "But your sorrow shall be turned into joy."   Now, upon the first day
of the week, very early in the morning, (Easter morning) the two Marys came to the sepulcher bringing the
sweet spices and ointments which they had prepared for
the purpose of anointing the body of the Lord Jesus,
but were greatly astonished when they saw that the
great stone, which they had seen rolled against the
door of the sepulcher on Friday afternoon, was rolled
away, and an angel sitting upon it whose countenace
was like lightning, and for fear of whom the keepers
did shake and become as dead men.   But to the women
he said, "Fear not ye ; for I know that ye seek Jesus
which was crucified, He is not here, for He is risen."

That you may know for a certainty that He is risen,
come and see the place where you saw Him laid only a
few hours ago. Now go quickly and tell His disciples
that He is risen from the dead, and behold He goeth
before you into Galilee, there shall ye see Him. And
they departed quickly from the sepulcher with fear and
great joy, and did run to carry the good news to the disciples. But how much greater their joy soon after their
departure when Jesus himself met them with the comfort-
ing words, '' Be not afraid, but go tell my brethren that
they go into Galilee, and there shall they see Me.''

The subject for conversation for the past two days
had been Jesus and the crucifixion, but now Jesus and
the resurrection. Some believed, but some doubted.
Others ran to the sepulcher and found it even as
the women had said. While the chief priests and
elders hired the soldiers to say that the disciples came
by night and stole him away while we slept. But how
should they know what had become of Him if they
were sleeping ? The truth was they were so overcome
with fear by the brightness of the angels' countenance
that they became as dead men, not knowing what was
transpiring around them. But it was truly good tidings
and great joy to the disciples of Christ on that Easter
morning.

> The rising God forsakes the tomb ;
> In vain the tomb forbids His rise ;
> Vain the stone, the watch, the seal,
> Christ has burst the gates of hell ;
> Death in vain forbids His rise ;
> Christ hath opened Paradise.

The spirit of Christ in the prophets had testified
beforehand the suffering of Christ and the glory that

should follow. That morning and that day was not only joyful to the disciples of Christ, but *glorious* ; it was "joy unspeakable and *full* of glory." Although 1863 years have rolled around since that Easter morning, yet we are as much interested in what then and there transpired as were the Marys, and Peter, and John, who were the first at the sepulcher, and who were permitted the same day to see their risen Lord, whom having not seen, we love ; in whom, though now we see Him not, yet believing, we [as well as they] rejoice " with joy unspeakable and full of glory."

Low Sunday. The first Sunday after Easter is so called because it was customary to repeat on this day some part of the solemnity which was used on Easter day, whence it took the name of Low Sunday, being celebrated as a feast, but of a lower degree than Easter day itself. The next Sunday after Easter has been popularly, so called in England, perhaps by corruption for close, (*Pascha Clausum*) close of Easter. *Dominica* in *Albis*, (the Sunday of white garments) a title anciently given to the first Sunday after Easter, because on this day those persons who had been baptized at Easter appear for the last time in the chrysomes, or white robes which they received at baptism. These were laid up in the church as evidences of their baptismal profession.

Rogation Days, (Latin *rogare*, to beseech,) are the Monday, Tuesday and Wednesday after Rogation Sunday and before Ascension Day, (Holy Thursday.) About the middle of the fifth century Mamertus, bishop of Vienna, upon the prospect of a great fire that threatened his diocese, appointed that extra-

ordinary prayer and supplication should be offered
up to God, with fasting, for averting the impending
evils upon the above mentioned days ; from which sup-
plications (called by the Latins *rogationes*) these days
have ever since been called Rogation Days. As retained
in our present calendar, they are simply private fasts.

f—Ascension Day, or Holy Thursday, one of the
great religious festivals of the Roman Catholic and Episcopal churches, is held on the fortieth day after Easter,
and ten days before Whitsun Day, to commemorate the
Ascension of our Lord into heaven. Ascension Day
has been observed at least since the year of our Lord
64, and perhaps earlier. Saint Augustine believed it
to have been instituted either by the apostles themselves, or the bishops immediately succeeding them.

Expectation Week is the whole of the interval between Ascension Day and Whitsun Day, so called
because at this time the apostles, according to the command of our Saviour, continued at Jerusalem, in earnest
prayer and expectation of the Holy Comforter which
was to abide with them forever. The Sunday between
Ascension Day and Whitsun Day is called Expectation
Sunday.

Pentecost, (Greek, *Pentecostos*, fiftieth), a Jewish festival ; so called because it was observed on the fiftieth
day after the feast of unleavened bread, called also the
feast of weeks, being celebrated seven weeks from the
feast of the Passover. It also commemorated the giving of the law on Mount Sinai upon that day. The
origin of the Anglo-Saxon name of White Sunday, which
also occurs in Icelandic, is somewhat obscure, for in
the Roman churches the *Dominica* in *Albis*, (Low Sun-

day, q v.) so called from the white robes then worn by
the persons baptized at Easter, has always been the
Sunday immediately following Easter. It hardly seems
probable that there should be another Sunday of White
Garments within six weeks. In German it is known
by the name *Pfingsten*, old German *Wingsten*, old English *Whitsun*, hence, probably, our word Whitsun Day,
not White Sunday.

g—Whitsun Day, or Pentecost, is the last of the four
great festivals in the ecclesiastical year, held in commemoration of the outpouring of the Holy Spirit on the
infant church ten days after the Ascension. Among
the last words of our Saviour to the apostles on the
very day that He was taken up, were " Behold I send
the promise of my Father upon you, but tarry ye in the
city of Jerusalem until ye be endowed with power from
on high." After ten days of earnest, believing prayer,
and expectation, suddenly, but not unexpectedly, there
came a sound from heaven as a rushing, mighty wind,
and it filled all the house where they were sitting, and
there appeared unto them cloven tongues like as of fire,
and it sat upon each of them ; and they were all filled
with the Holy Ghost, and began to speak with other
tongues as the Spirit gave them utterance ; and the
multitude came together, and were amazed, saying one
to the other what meaneth this ? Others mocking said,
these men are full of new wine. But Peter lifted up
his voice and said, these are not drunken as ye suppose, seeing it is but the third hour of the day, (nine
o'clock in the morning,) men are not usually drunk so
early in the morning ; but this is that which the prophet
Joel eight hundred years ago said should come to pass

in these last days ; the promise of the Father, the baptism of the Holy Ghost for which they had been waiting for the past ten days ; something of that *glory* that should follow the crucifixion, death, resurrection and ascension of the Lord Jesus into Heaven, the glory of the Christian church, of the Christian dispensation, and which is destined to fill the whole earth.

> " Waft, Waft, ye winds his story,
> And you, ye waters, roll,
> Till like a sea of glory,
> It spreads from pole to pole."

h—Trinity Sunday, the octave, or eighth day of the feast of Pentecost, is a church festival held in commemoration of the doctrine of the Trinity. The introduction of this day into the calendar is of comparatively recent date, it being established by Pope Benedict XI, in the year of our Lord 1305. It is probable that the zeal of many Christians against the use of images in the eighth and ninth centuries may have been the first cause of the appointment of a distinct day for meditating upon the nature of the Holy Trinity in unity, or the one true God as distinguished from idols. The reason for its late introduction is that in the creed of the church, and in its psalms, hymns, and doxologies, great prominence was given to this doctrine, and it was thought that there was no need to set apart a particular day for that which was done every day. This is the last of the movable feasts in the ecclesiastical year, being held eight weeks after Easter ; so it may happen as early as the 17th of May or as late as the 20th of June. The length of both Epiphany and Trinity-tide depend upon the date of Easter. As Epiphany-tide is short-

ened by the early date of Easter, so Trinity-tide is
lengthened proportionately, and as Epiphany-tide is
lengthened by the later date of Easter so Trinity-tide is
shortened proportionately ; so Trinity-tide may contain
no more than 196 days nor less than 161. (See tables
at the close of this chapter.)

All Saints Day, or All hallowmass (Anglo-Saxon *all*,
and *halig*, holy) a festival celebrated by the Roman
Catholic and Episcopal churches on the first day of
November, in honor of all Saints and martyrs.   It was
introduced into the Western church in the beginning
of the seventh century by Boniface.   The number of
saints being exceedingly multiplied, it was found too
burdensome to dedicate a feast day to each, there being,
indeed, scarcely hours enough in the year to distribute
among them all.   It was therefore resolved to commemorate on one day all who had no particular day.
By order of Gregory IV, it was celebrated on the first
of November, 834 ; formerly the first of May was the
day appointed.   It was introduced into England about
870, and is still observed in the English and Lutheran
churches, as well as the Church of Rome on the first
of November.

All-Souls' Day, a festival held by Roman Catholics
on the 2d of November, for special prayer in behalf of
all the faithful dead.   It was first introduced in 998,
by Odilon, Abbot of Clugni, who enjoined it on his
own order.   It was soon after adopted by neighboring
churches.   It is the day on which, in the Romish
church, extraordinary masses are repeated for the relief
of souls said to be in purgatory.   Formerly on this day
persons dressed in black perambulated the towns and

cities, each provided with a bell of dismal tone, which was rung in public places by way of exhortation to the people to remember the souls in purgatory. In some parts of the west of England it is still the custom for the village children to go around to all their neighbors souling, as they call it, collecting small contributions, and singing the following verses, taken down from two of the children themselves:

Soul! Soul! for a soul-cake,
Pray good mistress, for a soul-cake,
One for Peter, two for Paul,
Three for Him that made us all.

Soul! soul! for an apple or two;
If you've got no apples, pears will do,
Up with your kettle, and down with your pan,
Give me a good big one and I'll be gone.

The soul cake referred to in the verses is a sort of bun which the people used to make, and to give to one another on the 2d of November.

In the following tables there is presented at one view the day of the month on which the principal feasts and fasts fall in the ecclesiastical year 1817-18 and 1885-86. In the former Easter happens at its earliest possible date, in the latter at its latest date in this century:

YEAR 1817-18.

	Days in Each Season.	Sundays in Each Season.
a——Advent Sunday, November 30th; Advent-tide, . .	25	4
1st Ember Week, after December 13th; Ember Days, Wednesday, Friday and Saturday.		
b——Christmas, December 25th; Christmas-tide, ...	12	2
c———Epiphany, January 6th; Epiphany-tide, ...	12	1
Septuagesima Sunday, January 18th, ...	7	1
Paschal season from Jan. 18th to March 29th, 70 days		
Sexagesima Sunday, January 25th, ...	7	1
Quinquagesima Sunday, February 1st, ...	3	1
Shrove-tide, (confession time) Shrove Tues., Feb. 3d.		
d——Ash Wednesday, Feb 4th, Lent begins; Lenten-tide,	45	6
First Sunday in Lent (Quadragesima) February 8th.		
2d Ember Week after first Sunday in Lent; Ember Days, Wednesday, Friday and Saturday.		
Holy Week, the week before Easter; Special Days, Palm Sunday, Spy Wednesday, Maundy Thursday, and Good Friday, March 15th, 18th, 19th and 20th.		
e——Easter Sunday, March 22d; Easter-tide, ...	39	6
Low Sunday, March 29th; Paschal Season ends.		
Rogation Sunday, April 26th; Rogation Days, the Monday, Tuesday and Wednesday after Rogation Sunday.		
f——Ascension Day (Holy Thursday), April 30th; Ascension-tide, ...	10	1
Expectation Sunday, First Sunday after Ascension, May 3d.		
g——Whitsun Day, (Pentecost) May 10th; Whitsun-tide, .	7	1
3d Ember Week, after Whitsun Day; Ember Days, Wednesday, Friday and Saturday.		
h——Trinity Sunday, May 17th; Trinity-tide, ...	196	28
4th Ember Week, after September 14th; Ember Days, Wednesday, Friday and Saturday.		
All Saints' Day, November 1st.		
All Souls' Day, November 2d.	——	
Appendix K.	364	52

123

YEAR 1885–86.

	Days in Each Season.	Sundays in Each Season.
a——Advent Sunday, November 29th, Advent-tide, . . .	26	4
1st Ember Week, after December 13th; Ember Days, Wednesday, Friday and Saturday.		
b——Christmas, December 25th; Christmas-tide, ...	12	2
c——Epiphany, January 6th; Epiphany-tide, ...	46	6
Septuagesima Sunday, February 21st, ...	7	1
Paschal season, from Feb. 21st to May 2d, 70 days.		
Sexagesima Sunday, February 28th, ...	7	1
Quinquagesima Sunday, March 7th, ...	3	1
Shrove-tide, (confession time) Shrove Tues., Mar. 9th.		
d——Ash Wednesday, March 10th, Lent begins; Lenten-tide	46	6
First Sunday in Lent (Quadragesima) March 14th.		
2d Ember Week after first Sunday in Lent; Ember Days, Wednesday, Friday and Saturday.		
Holy Week, the week before Easter; Special Days, Palm Sunday, Spy Wednesday, Maundy Thursday and Good Friday, April 18th, 21st, 22d and 23d.		
e——Easter Sunday, April 25th; Easter-tide, . .	39	6
Low Sunday, May 2d, Paschal Season ends.		
Rogation Sunday, May 30th; Rogation Days, the Monday, Tuesday and Wednesday after Rogation Sunday.		
f——Ascension Day (Holy Thursday), June 3d; Ascension-tide, ...	10	1
Expectation Sunday, first Sunday after Ascension, June 6th.		
g——Whitsun Day, (Pentecost) June 13th; Whitsun-tide,	7	1
3d Ember Week, after Whitsun Day; Ember Days, Wednesday, Friday and Saturday.		
h——Trinity Sunday, June 20th; Trinity-tide, ...	161	23
4th Ember Week, after September 14th; Ember Days, Wednesday, Friday and Saturday.		
All Saints' Day, November 1st.		
All Souls' Day, November 2d		
Appendix K.	364	52

CHAPTER VIII.

HEBREW CALENDAR.

To the Bible student a knowledge of the Hebrew calendar is indispensable, if he would know how the date
of events recorded in the Bible are made to correspond
with our present English calendar. From the exodus
(1491 B. C.) downward, the Hebrew month was lunar,
and commenced invariably with the new moon.

Dr. Smith, author of Bible Dictionary, says that
the terms for month and moon have the same close con-
nection in the Hebrew language as in our own, only
the Hebrew Codesh (that is new moon) is, perhaps,
more distinctive than the corresponding term in our
language; for it expresses not simply the idea of a
lunation, but the recurrence of a period commencing
definitely with the new moon. Though the identifica-
tion of the Jewish month with our own cannot be effected with precision on account of the variations that
must necessarily exist between the lunar and the solar
month, each of the former ranging over portions of two
of the latter, still it can be shown how they may be
made to coincide very nearly by a systematic method
of intercalation.

Now from new moon to new moon again, is about
29½ days; therefore, the Hebrew year consisted of
354 days, for 29½ × 12 = 354; so that the epact, (which
is the excess of the solar year beyond the lunar) is
eleven days. Hence, had they no method of intercala-

tion, the commencement of their year would go back eleven days every year, and consequently make a revolution of the seasons every thirty-three years, for 365 ÷ 11 = 33 nearly.

To illustrate, let us suppose that the new moon of Nisan, which is the first month in the Sacred year, should on any given year fall on the 10th of April, then the following year it would fall on the 30th of March, which is eleven days earlier; the second year it would fall on the 19th of March or twenty-two days earlier; the third year the new moon would fall on the 8th of March or thirty-three days earlier, but that would not be the new moon of Nisan, which cannot happen earlier than the 11th, so the following moon which happens thirty days later on the 7th of April is the new moon of Nisan. Hence it may be seen that by intercalating a full month every three years, or which comes nearer to accuracy seven times in nineteen years, restores the coincidence of the solar and the lunar year, and consequently the moons to the same day of the month on which they fell nineteen years before.

The method of designating the months previous to the exodus, was by their numerical order, as the ancient Hebrews had no particular name to express their month. They said the first, second and third month, and so on. No names of months appear in the Bible until about the time of the institution of the passover, when the Lord spake unto Moses and Aaron in the land of Egypt, saying this month, (Abib, which appears to have had its origin in Egypt,) shall be unto you the beginning of months; it shall be the first month of the year to you.

The names of the months appear to belong to two distinct periods. In the first place we have those peculiar to the Jews previous to the captivity, viz : Abib, the first month in commemoration of the exodus ; Zif, the second, Ethanim, the seventh, and Bul, the eighth. These names are of Hebrew origin, and have reference to the characteristics of the seasons, a circumstance which clearly shows that the months, by intercalation, were made to return at the same period of the year. Thus, Abib was the month of the ears of corn, that is the month in which the ears of corn became full, or ripe on the 16th day, that is the 2d day of the feast of unleavened bread. Zif, the month of blossoms or the bloom of flowers. Ethanim, the month of gifts, that is of fruits, and Bul, the month of rain. These were superceded after the captivity, by Nisan, Iyar, Tisri and Hesvan, or Marchesvan.

Marchesvan, coinciding as it does with the rainy season in Palestine, is considered a pure Hebrew term. The modern Jews consider it a compound word, from mar, drop, and Chesvan ; the former betokening that it was wet, and the latter being the proper name of the month. Hence the name indicates the wet month. In the second place we have the names of six others which appear in the Bible subsequently to the Babylonian captivity, viz.: Sivan, the third ; Elul, the sixth ; Kislev, the ninth ; Tebet, the tenth ; Sebat, the eleventh, and Adar, the twelfth. There are two other months whose names do not appear in the Bible, viz.: Tamuz, the fourth, and Ab, the fifth. The name of the intercalary month is called Ve-Adar, or 2d Adar because placed in the calendar after Adar and before Nisan.

Dr. Smith says these names are probably borrowed from the Syrians in whose regular calendar we find names answering to most of them. He also says it was the opinion of the Talmudists, that these names were introduced by the Jews who returned from the Babylonish captivity, and also that they are certainly used exclusively by writers of the post-Babylonian period.

Inasmuch as the Hebrew months coincided with the seasons, as we have already shown, it follows as a matter of course, that an additional month must have been inserted every third year, which would bring the number up to thirteen. No notice, however, is taken of this month in the Bible, neither have we reason to think that it was inserted according to any exact rule, but it was added whenever it was discovered that the barley harvest did not coincide with the ordinary return of the month Abib. It has already been shown that in the modern Jewish calendar the intercalary month is introduced seven times in nineteen years, according to the Metonic, or lunar cycle which was adopted by the Jews about 360 A. D.

The Hebrew calendar is dated from the creation, which is supposed to have taken place 3761 years before Christ. Hence, to find the number of cycles elapsed since the creation, also the number in the cycle, we have the following rule: Add 3761 to the date, divide the sum by nineteen; the quotient is the number of cycles, and the remainder is the number in the cycle. Should there be no remainder, the proposed year is, of course, the last or nineteenth of the cycle. Thus, for the year 1883, we have $1883+3761 \div 19 =$ 297, remainder 1; therefore, 297 is the number of cycles, and 1 the number in the cycle. Again, for the year 1893, we have $1893+3761 \div 19 = 297$, remainder 11 ; therefore 297 is the number of cycles, and 11 the number in the cycle. Again for the year 1901, we have $1901+3761 \div 19 = 298$, remainder 0 ; therefore 298 is the number of cycles, and 19 the last of the cycle. Hence it may be seen that the present cycle commenced with 1883, that 11 is the number in the cycle for the present year 1893, also that the cycle ends with 1901 ; so that the next cycle commences with 1902. If the remainder after dividing by nineteen be 3, 6, 8, 11, 14, 17 or 19 (0), the year is intercalary or embolismic, consisting of 384 days ; if otherwise it is ordinary, containing only 354 days ; so that in a cycle of nineteen years, we have twelve ordinary years of 354 days each, and seven embolismic years of 384 days each. But, in either case, the year is sometimes made a day more, and sometimes a day less, in order that certain festivals may fall on proper days of the week for their due observance. Hence the ordinary year may consist of 353, 354 or 355 days, and the embolismic year of 383, 384 or 385 days.

In the modern Jewish calendar the New Year commences with the new moon of Tisri, which may happen as early as the 5th of September or as late as the 5th of October. The new moon of Nisan, which is the first month in the Sacred year, may happen as early as the 11th of March or as late as the 11th of April. It should be borne in mind that the names of the months Abib, Zif, Ethanim and Bul were superceded after the captivity, by Nisan, Iyar, Tisri and Hesvan or Marchesvan ; also the name of the third month in the civil year, Chisleu in the Bible, Kislev in the modern Jewish calendar. In table No. 1 we have the names of the months in numerical order, also the number of days in each month. Though the months consist of 30 and 29 days alternately, yet, in the embolismic year, Adar, which in common years has 29 days, is given 30 days, and 2d Adar 29 ; so that two months of 30 days come together. Table No. 2 shows the earliest and the latest possible date of the new moons of each of the months respectively.

TABLE I.   HEBREW MONTHS.

Sacred Year.			Civil Year.			
Nisan	-	30	Tisri	-	-	30
Iyar	-	29	Hesvan	-	-	29
Sivan	-	30	Kislev	-	-	30
Tamuz	-	29	Tebet	-	-	29
Ab	-	30	Sebat	-	-	30
Elul	-	29	Adar	-	-	30
Tisri	-	30	2d Adar, Embolismic			29
Hesvan	-	29	Nisan	-	-	30
Kislev	-	30	Iyar	-	-	29
Tebet	-	29	Sivan	-	-	30
Sebat	-	30	Tamuz	-	-	29
Adar	-	30	Ab	-	-	30
2d Adar, Embolismic		29	Elul	-	-	29

TABLE II.   HEBREW MONTHS.

Nisan,	March 11th	or April 11th
Iyar,	April 11th	" May 10th
Sivan,	May 10th	" June 9th
Tamuz,	June 9th	" July 9th
Ab,	July 9th	" August 7th
Elul,	August 7th	" September 5th

Tisri,	September 5th	or October 5th
Hesvan,	October 6th	" November 4th
Kislev,	November 4th	" December 3d
Tebet,	December 3d	" January 2d
Sebat,	January 3d	" February 10th
Adar,	February 10th	" March 12th

The charts on the three following pages are used to illustrate the correspondence of the Hebrew months with our own. Each chart represents the ecliptic, which is the apparent path of the Sun or real path of the Earth, also the names of the months as they occur in their seasons. The figures represent the days of the month on which the new moons of the Hebrew calendar fall. These charts represent the month and the day of the month on which both the Sacred and the Civil year begins and ends for three successive years. Hence it may be seen that by intercalating a month every three years the new moons are restored, very nearly, to the place they occupied three years before.

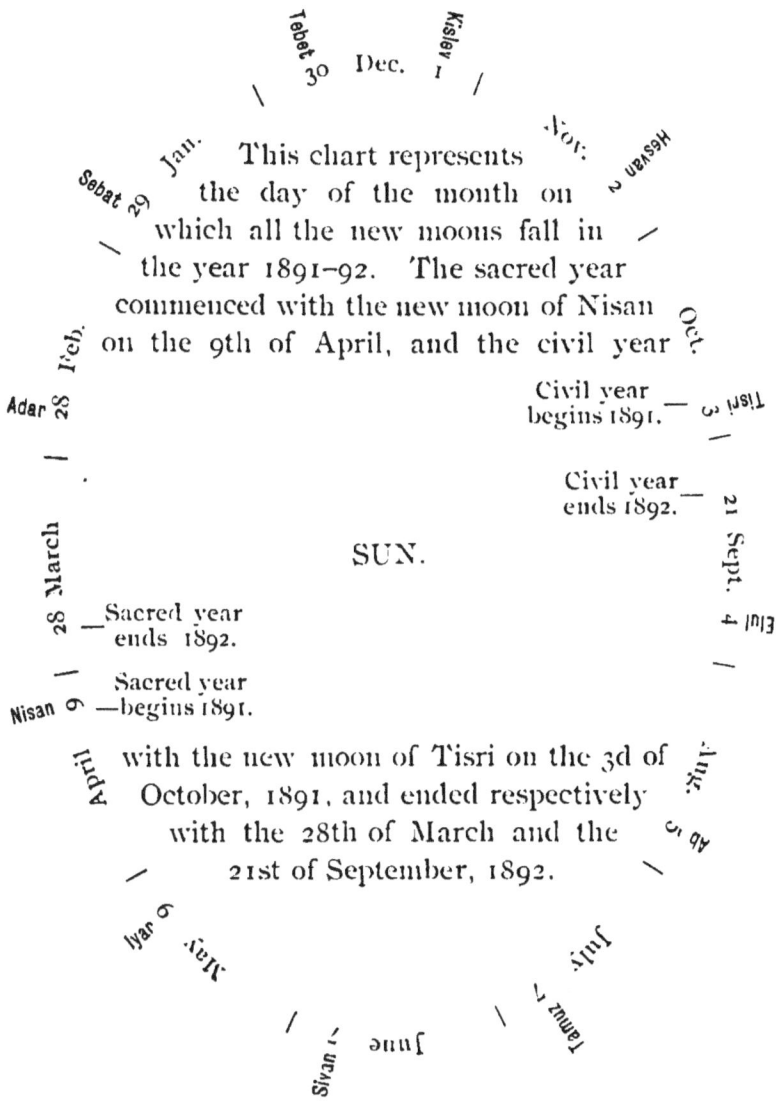

CHART I

Tebbet 30 Dec. Kislev 1

Jan. This chart represents Nov. Heshvan

Sebat 29 the day of the month on

which all the new moons fall in

the year 1891–92. The sacred year

commenced with the new moon of Nisan Oct.

Feb. on the 9th of April, and the civil year

Adar 28

Civil year __
begins 1891. Tisri 3

Civil year __
ends 1892. 21 Sept.

28 March

SUN.

__Sacred year
ends 1892. 4 Elul

Sacred year
Nisan 9 —begins 1891.

April with the new moon of Tisri on the 3d of Aug.

October, 1891, and ended respectively

with the 28th of March and the 5 Ab

21st of September, 1892.

Iyar 9

May July

Sivan 1 June Tammuz 7

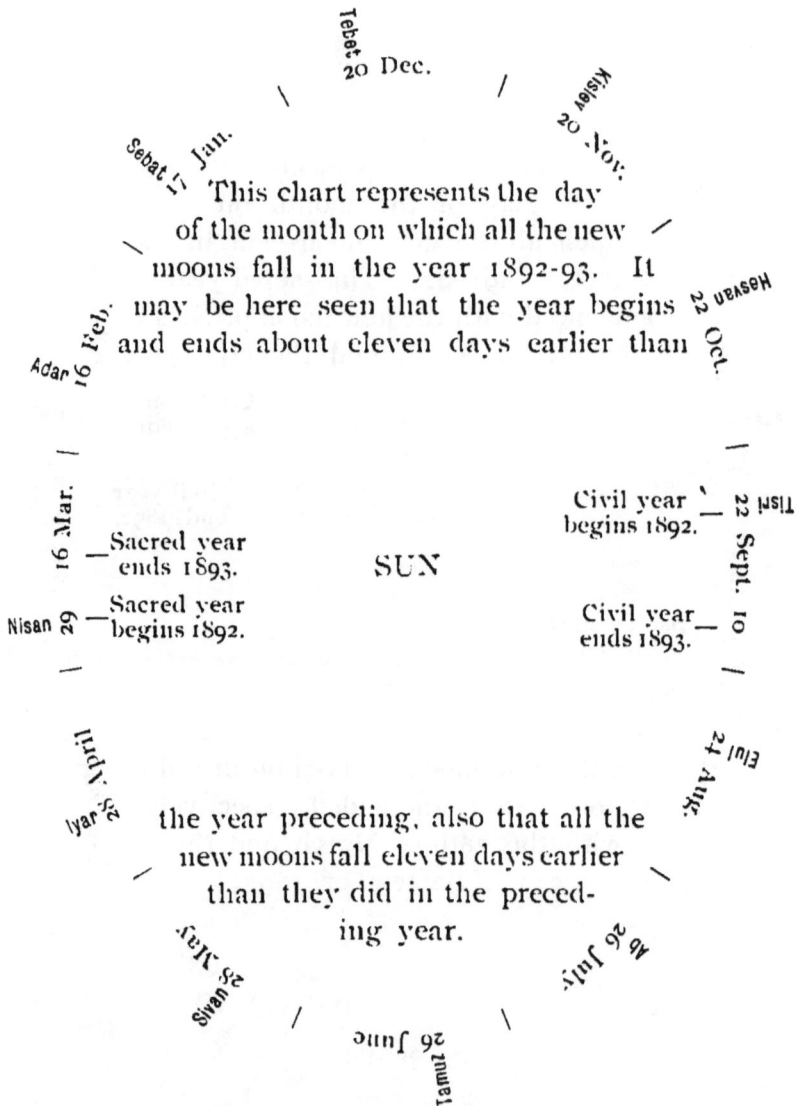

CHART II.

Tebet 20 Dec.

Sebat 17 Jan.

Kislev 20 Nov.

Adar 16 Feb.

Hesvan 22 Oct.

This chart represents the day of the month on which all the new moons fall in the year 1892-93. It may be here seen that the year begins and ends about eleven days earlier than

16 Mar.

Tisri 22 Sept.

— Sacred year ends 1893.

Civil year begins 1892.

SUN

Nisan 29

— Sacred year begins 1892.

Civil year ends 1893.

10

Iyar 28 April

Elul 24 Aug.

the year preceding, also that all the new moons fall eleven days earlier than they did in the preceding year.

Sivan 25 May

Ab 26 July

Tamuz 26 June

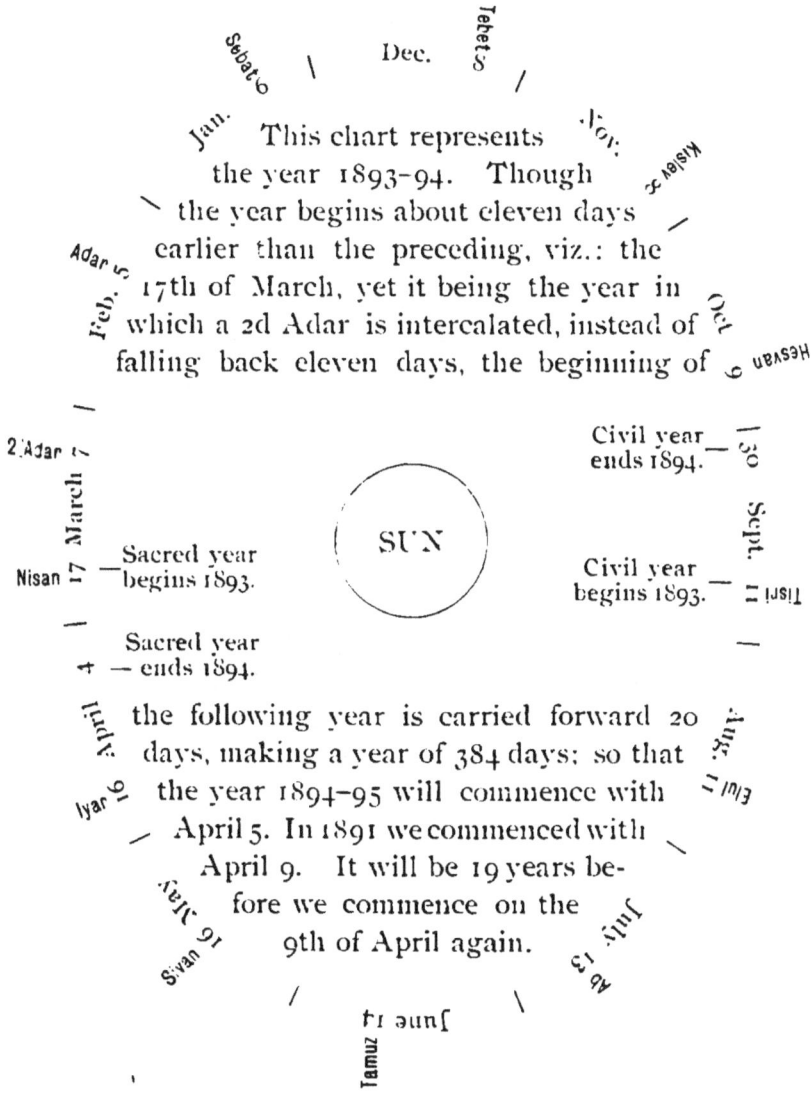

133

CHART III.

Tebeth 3

Sebat 6   Dec.

Jan.   This chart represents   Nov.

Kislev 8

Adar 5   the year 1893-94. Though

the year begins about eleven days

earlier than the preceding, viz.: the

Feb.   17th of March, yet it being the year in

Oct.

which a 2d Adar is intercalated, instead of

falling back eleven days, the beginning of

Heshvan 9

2.Adar 7

Civil year ends 1894. 30

March 17

Sacred year begins 1893.

SUN

Civil year begins 1893.

Sept.

Nisan 17

Tisri 1

Sacred year ends 1894.

April 4

the following year is carried forward 20

days, making a year of 384 days; so that

the year 1894-95 will commence with

Aug.

Elul 3

Iyar 6

April 5. In 1891 we commenced with

April 9. It will be 19 years be-

May 7   fore we commence on the

Sivan 91   9th of April again.

July 13

Ab 4

June 14

Tamuz

APPENDIX.

A.—PAGE 12.

Authors differ in regard to the length of the solar year. One gives 365 days, 5 hours, 47 minutes and 51.5 seconds; another, 365 days, 5 hours, 48 minutes and 46 seconds; and still another, 365 days, 5 hours, 48 minutes and 49.62 seconds. In this work the last has been accepted as the true length of the solar year, and all calculations have been made accordingly.

B.—PAGE 19.

There is an apparent discrepancy among authors in regard to the intercalary day. While one asserts that it was between the 24th and 25th of February, another equally reliable, says that the 25th was the sexto calendas and the 24th was the bis-sexto calendas of the Julian calendar. Now it should be borne in mind that the Julian calendar is the basis of our own, and is identical with it in the number of months in the year, and in the number of days in the month. Also when the method of numbering the days from the beginning of the month was adopted, the intercalation was made to correspond with the intercalary day in the Julian calendar.

As in the Julian calendar there were twice the sixth day, so in the reformed calendar there were twice the 24th day, which was equivalent to 29 days in February. When the calendar was again corrected, making the 29th the intercalary day, then the 24th corresponded with the bis-sexto calendas of the Julian calendar. This reconciles the apparent discrepancy. While one author refers to the calendar in which the Julian rule of intercalation is adopted, another refers to the calendar when so corrected as to make the 29th of February the intercalary day. See following table :

	JULIAN METHOD OF INTERCALATION.			MODERN METHOD OF INTERCALATION.	
1	*Cal.*	Calendae	1	*Cal.*	Calendae
2	4	Quarto Nonas	2	4	Quarto Nonas
3	3	Tertio Nonas	3	3	Tertio Nonas
4	2	Pridie Nonas	4	2	Pridie Nonas
5	*Nomes*	Nonae	5	*Nomes*	Nonae
6	8	Octavo Idus	6	8	Octavo Idus
7	7	Septimo Idus	7	7	Septimo Idus
8	6	Sexto Idus	8	6	Sexto Idus
9	5	Quinto Idus	9	5	Quinto Idus
10	4	Quarto Idus	10	4	Quarto Idus
11	3	Tertio Idus	11	3	Tertio Idus
12	2	Pridie Idus	12	2	Pridie Idus
13	*Ides*	Idus	13	*Ides*	Idus
14	16	Sextodecimo Calendas	14	16	Sextodecimo Calendas
15	15	Quintodecimo Calendas	15	15	Quintodecimo Calendas
16	14	Quartodecimo Calendas	16	14	Quartodecimo Calendas
17	13	Tertiodecimo Calendas	17	13	Tertiodecimo Calendas
18	12	Duodecimo Calendas	18	12	Duodecimo Calendas
19	11	Undecimo Calendas	19	11	Undecimo Calendas
20	10	Decimo Calendas	20	10	Decimo Calendas
21	9	Nono Calendas	21	9	Nono Calendas
22	8	Octavo Calendas	22	8	Octavo Calendas
23	7	Septimo Calendas	23	7	Septimo Calendas
24	6	Bis-Sexto Calendas	24	6	Bis-sexto Calendas
24	6	Sexto Calendas	25	6	Sexto Calendas
25	5	Quinto Calendas	26	5	Quinto Calendas
26	4	Quarto Calendas	27	4	Quarto Calendas
27	3	Tertio Calendas	28	3	Tertio Calendas
28	2	Pridie Calendas	29	2	Pridie Calendas

C.—PAGE 20.

The city where the great council was convened in 325 is not in France, as some have supposed, that being a more modern city of the same orthography, but pronounced Nees. The city which is so frequently referred to in this work is in Bythinia, one of the provinces of Asia Minor, situated about 54 miles southeast of Constantinople, of the same orthography as the former, but pronounced Ni'ce, and was so named by Lysimachus, a Greek general, about 300 years before Christ, in honor of his wife Nicea.

D.—PAGE 23.

Between the 23d and 24th of February, 46 years before Christ, there was intercalated a month of 23 days according to an established method, but still the civil year was in advance of the solar year by 67 days; so that when the Earth in her annual revolutions should arrive to that point of the ecliptic marked the 22d of October, it would be the 1st day of January in the Roman year.

Cæsar and his astronomers, knowing this fact and fixing on the 1st day of January, 45 years before Christ and 709 from the foundation of Rome, for the reformed calendar to take effect, were under the necessity of intercalating two months, together consisting of 67 days. Now, as the civil year would end on the 22d of October, true or solar time, it would be reckoned in the old calendar the 1st day of January; so they let the old calendar come to a stand while the Earth performs 67 diurnal revolutions, and thereby restored the concurrence of the solar and the civil year.

As an illustration, let us suppose that in a certain shop where hangs a regulator are two clocks to be regulated. Both are set with the regulator at 8 a. m. to see how they will run for ten consecutive hours. It was found that when it was 6 p. m., by the first clock, it was 5:50 by the regulator, the clock having gained one minute every hour.

To rectify this discrepancy we must intercalate 10 minutes by stopping the clock until it is 6 by the regulator. By this means the coincidence is restored, and the time lost in the preceding hours is now reckoned in this last hour, making it to consist of 70 minutes. By this it may be seen how Cæsar reformed the Roman calendar. The Roman year was too short, by reason of which the calendar was thrown into confusion, being 90 days in advance of the true time, so that December, January and February took the place in the seasons of September, October and November, and September, October and November the place of June, July and August. To make the correction he must stop the old Roman clock (the calendar) while the Earth performs 90 diurnal revolutions to restore the concurrence of the solar and the civil year, making the year 46 B. C. to consist of 445 days.

It was also found that when it was 6 p. m., by the regulator, it was only 5:50 by the second clock, it having lost one minute every hour. To rectify this discrepancy we must suppress 10 minutes, calling it 6 p. m., turning the hands of the clock to coincide with the regulator, making the last hour to consist of only 50 minutes, too much time having been reckoned in the preceding hours. It may be seen by this illustration, how Gregory corrected the Julian calendar, the Julian year was too long, consequently behind true or solar time, so that when the correction was made in 1582, the ten days gained had to be suppressed to restore the coincidence, making the year to consist of only 355 days.

As the solar year consists of 365 days and a fraction, Cæsar intended to retain the concurrence of the solar and the civil year by intercalating a day every four years; but this made the year a little too long, by reason of which it became necessary, in 1582, to rectify the error, and by adopting the Gregorian rule, three intercalations are suppressed every 400 years; so that by a series of intercalations and suppressions, our calendar may be preserved in its present state of perfection.

E.—PAGE 23.

As the day and the civil year always commence at the same instance, so they must end at the same instance; and as the solar year always ends with a fraction, not only of a day, but of an hour, a minute and even a second; so there is no rule of intercalation by which the solar and the civil year can be made to coincide exactly. But the discrepancy is only a few hours in a hundred years, and that is so corrected by the Gregorian rule of intercalation that it would amount to a little more than a day in 4,000 years; and by the improved method less than a day in 100,000 years.

F.—PAGE 26.

It has been stated that by adopting the Julian rule of intercalation, time was gained; it has also been stated that by the same rule time was lost. Now both are true. Time is gained in that there is too much time in a given year, in other words, the year is too long ; but what is gained in a given year is lost to the following year.

As an illustration let us take the case of the supposed solar year of 365 days, and the civil year of 366. The civil year would gain one day every year, or be too long by one day ; but the one day gained is lost to the following years, and if continued 31 years, when the Earth is in that part of its orbit marked the 1st day of January 32, the civil year would reckon the 1st day of December 31 ; so that in the thirty-one years would reckon thirty-one days too much, and before the civil year is completed, the Earth will have passed on in its orbit to a point marked the 1st day of February.

Now to reform such a calendar, we would have to suppress or drop the thirty-one days, by calling the 1st day of December the 1st day of January, and thus the month of December would disappear from the calendar in the year 31, making a year of only eleven months, consisting of 334 days.

If this method be continued 92 years, there would be gained 92 days, to the loss of 92 days in the year 92. If the calendar be now reformed by suppressing 92 days, calling the 1st day of October, 92, the 1st day of January, 93, then October, November and December would disappear from the calendar in the year 92 ; and if continued 365 years there would be crowded into 364 years, 364 days too much ; gained to the 364 years to the total loss of the year 365, passing from 364 to 366 ; 365 disappearing from the calendar.

G.—PAGE 50.

An era is a fixed point of time from which a series of years is reckoned. Among the nations of the Earth there are no less than twenty-five different eras ; but the most of them are not of enough importance to be mentioned here. Attention is particularly called to the Roman era which commenced with the building of the city of Rome 753 years before Christ.

Also the Mahometan era, or the era of the Hegira, employed in Turkey, Persia and Arabia, which is dated from the flight of Mahomet from Mecca to Medina, which was Thursday night, the 15th of July, A. D., 622, and it commenced on Friday, the day following.

But there is a point from which all computation originally commenced, namely, the creation of man. Such an era is called the Mundane era. Now there are different Mundane eras—the common Mundane era 4,004 B. C., the Grecian Mundane era 5,598 B. C., and the Jewish Mundane era 3,761 B. C. All these commence computation from the same point, but differ in regard to the time which has elapsed since their computation commenced. God's people used the Mundane era, until the Great Creator appeared among us, as one of us, in the person of our Lord Jesus Christ, to accomplish the great work of redemption ; then His name was introduced as the turning point of the ages, the starting point of computation.

This was done by Dionysius Exiguus in the year of our Lord about 540, known at that time as the Dionysian, as well as the Christian era, and was first used in historical works by the venerable Bede early in the eighth century. "It was a great thought of the little monk(whether so called from his humility or littleness of stature is unknown), to view Christ as the turning point of the ages, and to introduce this view into chronology."

All honor to him who introduced it, and to the nations which have approved, for thus honoring the Great Redeemer. Dionysius probably did not know, neither is it now known for a certainty the year of Christ's birth, but it is evident, however, from the best authorities, that the era commenced at least five years too late, and probably more.

H.—PAGE 57.

It is recorded that, in the time of Numa, the vernal equinox fell on the 25th of March, and that Julius Cæsar restored it to the 25th, when he reformed the ancient Roman calendar in the year 46 B. C. It is also recorded that in less than 400 years from that time, at the meeting of the Council of Nice in 325, it had fallen back to the 21st—four days in less than 400 years.

Now there is an error somewhere, for it is found by actual computation that the discrepancy between the solar and the Julian year is about three days in 400 years. It certainly is true that the vernal equinox fell on the 21st in 325, and was restored to that place by Gregory in 1582 ; since which time it has been made to fall on the 21st by the Gregorian rule of intercalation. Again it is stated by the same author that the discrepancies in time from Cæsar to Gregory is thirteen days, from the Council of Nice to Gregory ten days ; now $10+4=14$. While our author states it is thirteen days, he also states it is fourteen days ; a discrepancy of one day. The mistake evidently is in making the 25th instead of the 24th, the date of the vernal equinox in the time of Cæsar, consequently a difference of four days instead of three from Cæsar to the Council of Nice.

I.—PAGE 59.

The concurrence of the solar and the civil year was restored by Gregory in 1582, or 1600 is the same in computation ; but the discrepancy between civil and solar time is 11 minutes and 10.38 seconds every year, which in 100 years will amount to 18 hours, and 37.3 minutes ; reckoned in round numbers 18 hours, and is represented on the chart, hours behind time 18.

The intercalary day or 24 hours being suppressed in 1700, causes the civil year to be 6 hours in advance of the solar, and is represented on the chart 6 hours in advance.

Now this discrepancy of 18 hours for the next 100 years, will cause the civil year in 1800 to be 12 hours behind ; again suppressing the intercalation it will be 12 hours in advance. In 1900 it will be 6 hours behind, but the correction makes 18 hours in advance. The 18 hours gained the next 100 years restores the coincidence in the year 2000 and so on, the solar and the civil year being made to coincide very nearly every 400 years.

From close examination it will become evident that the solar and the civil year coincide twice every 400 years, though no account is made of it in computation. From 6 hours in advance in 1700, the civil year falls back to 12 hours behind the solar in 1800, consequently they must coincide in 1733.

Again from 12 hours in advance in 1800, it falls back to 6 hours behind the solar in 1900, consequently they must coincide again in 1867.

Discrepancy between Julian and solar time in —1 year is (365d. 6h.)—(365d. 5h. 48m. 49.62s.)=(11m. 10.38s.)

100	years is (11m. 10.38s.) × 100	= (18h. 37.3.)
400	" (18h. 37.3m.) × 4	= (3d. 2h. 29.2m.)
4,000	" (3d. 2h. 29.2m.) × 10	= 31d. 0h. 52m.)
100,000	" (31d. 0h. 52m.) × 25	= (773d. 21h. 40m.)

Discrepancy between Gregorian and solar time in—

1	year is	-	-	-	-	.373m.
100	years is .373m. × 100 =	-	-	-		37.3m.
400	years 37.3m. × 4 =	-	-	-	2h. 29.2m.
4,000	" (2h. 29.2m.) × 10 =		-	1d. 0h. 52m.
100,000	" (1d. 0h. 52m.) × 25 =	-	25d. 21h. 40m.

Discrepancy between corrected Gregorian and solar time in—

4,000 years is (1d. 0h. 52m.)—1 day=	-		52m.
100,000 " " (52m. × 25)=	-		21h. 40m.

J.—PAGE 89.

Lilius, author of the "Extended Table of Epacts," says, when the full moon falls on the 20th of March, the following moon, which happens 29 days later, is the paschal moon, making the 18th of April its latest possible date. For, says he, because of the double epact that occurs on the 4th and 5th of April that lunation has only 29 days. It may have been very convenient for Lilius, in his peculiar method of determining the date of the paschal moon, to give to that lunation only 29 days; but nevertheless, when he did so, it was at the expense of accuracy, for he makes a difference of

12 days in the date of the paschal moon of that year, and the year preceding, and only 10 days difference between that year and the succeeding year ; whereas the difference is uniformly 11 days from year to year through the whole cycle of 19 years.

By referring to the table on the 93d page, it will be seen that, in fixing the date of the paschal moon, six times in a cycle of 19 years the full moon falls before the 21st of March, and in every instance except this one the following moon is reckoned by Lilius 30 days later. By this uniform method of determining the date of the paschal moon, we make the 19th of April instead of the 18th, its latest possible date ; so it should be borne in mind that whenever the 19th of April is the date of the paschal moon, as indicated in the tables commencing with the 93d page, that Lilius, and probably most, if not all other authors, have the 18th.

Now it is admitted that notwithstanding the cumbersome apparatus employed by Lilius in his calculations, the conditions of the problem are not always satisfied, nor is it possible that they can be always satisfied by any similar method of proceeding. We admit that none of these calculations are perfectly exact, but the sum of the solar and lunar inequalities is compensated in the whole period, or corrections made at the end of certain periods, not by interrupting the order of a uniform method during the cycle of 19 years.

Now the table of epacts was introduced by Lilius himself, making the excess of the solar year beyond the lunar, in round numbers 11 days. Then why interrupt this order every 19 years, for a period of 114 years ; that is from 1596 to 1710, by making the epact

12 days for one year, and the following year only 10? After which, from 1710 to 1900, a period of 190 years, according to Lilius' own calculations, the epact is uniformly 11 days, coinciding exactly with the calculations made in this work.

Then again after the year 1900, he gives to that particular lunation, in every lunar cycle for a period of 304 years, only 29 days; and having done so, he is under the necessity of giving only 29 days to another lunation in the same cycle, and also to all the cycles in the period to avoid the absurdity of making the paschal moon fall twice on the same day in the course of a lunar cycle.

By reference to the 101st page, opposite the year 1905, it will be seen that the date of the paschal moon is the 19th of April. Lilius, by giving to that lunation only 29 days, makes its date the 18th; and then again in the year 1916, lest he should make the paschal moon fall twice on the 18th of April in the course of a lunar cycle, (a thing which cannot really occur) he for the first time in more than 400 years, gives only 29 days to a second lunation in the same cycle and of course to all the cycles in the period of 304 years. Now the epacts for a lunar cycle of 19 years are represented thus:

$$0, \quad 11, \quad 22, \quad 3, \quad 14, \quad \overset{26}{\overline{25}}, \quad 6, \quad 17, \quad 28, \quad 9, \quad 20$$

$$1, \quad 12, \quad 23, \quad 4, \quad 15, \quad \overset{27}{\overline{26}}, \quad 7, \quad 18$$

The number 26 placed over the 25 shows Lilius' first error in giving to that lunation only 29 days. He thereby makes a difference of 12 days between the epact 14 and 26, and only 10 between 26 and 6. He now has two epacts of the same number 26. In order to get out of the dilemma he makes that 27, by giving to another lunation only 29 days.

K.—PAGE 122–3.

It will probably be noticed that according to the showing in the tables the ecclesiastical year contains only 364 days. The reason for this is, that Advent Sunday, which is the first day of the year, happens one day earlier every year until it occurs on the 27th of November, its earliest possible date; then the first Sunday after the 26th of November, which is Advent Sunday, falls on the 3d of December, its latest possible date, so that the year begins six days later, making a year of 371 days. Then there is the loss of a day every year until Advent Sunday again falls on the 27th of November and so on. Hence, did the civil year always consist of 365 days, then the ecclesiastical year would always contain either 364 or 371 days. But as every fourth year contains 366 days, this order is so interrupted that sometimes the first Sunday falls on the 2d instead of the 3d of December; so that the year begins only five days later, making a year of only 370 days. Hence the ecclesiastical year may consist of either 364, 370 or 371 days. But five times out of six it will contain only 364 days.

L.—PAGE 83.

But why did the Pope, in correcting the Julian calendar in 1582, not correct the whole error of thirteen
days? Why did he leave the three days uncorrected?
This question has been asked an hundred times, but a
correct answer has never yet been given. Some say
that the Pope did according to his best ability, and
would make us believe that neither he nor his astronomers knew what the error was. This is not true, for
history records the fact of the error, and just what that
error was. He simply did not want to correct the three
days, and for good reasons, which we shall endeavor to
show ; reasons which every churchman ought to know.

When Cæsar formed his calendar, 46 B. C., the vernal equinox fell on the 24th of March. At the meeting of the Council of Nice, in 325, it had fallen back to
the 21st, the error being three days in about 400 years.
Now it should be borne in mind that the Julian calendar was the only one in use at that time, and for the
next 1257 years, when in 1582, it was corrected by
Pope Gregory XIII. Easter, and all the movable
feasts, had been unsettled during the 1257 years intervening, from the Council of Nice to Gregory, on account of the errors of the Julian calendar. The Easter
question had been the cause of a good deal of discussion between the Eastern and Western churches during the second and third centuries, as they could not
agree on the day of the week on which that event
should be celebrated.

The Western churches observed the nearest Sunday
to the full moon of Nisan. The Asiatics, on the other
hand, adopted the 14th of Nisan upon which to com-

memorate the crucifixion, and observed the festival of
Easter on the third day following, upon whatever day
of the week that might fall. Finally, the Council of
Nice was convened, and the matter came before that
council, and a reconciliation was accomplished. It was
then and there agreed by the two parties that Easter
should be celebrated on the first Sunday after the full
moon that falls upon or next following the day of the
vernal equinox, and that the 21st of March should be
accounted the day of the vernal equinox.

It has already been shown that the error in the Julian
calendar is three days in 400 years ; so that in 400
years from the Council of Nice the vernal equinox had
fallen back to the 18th of March ; in 800 years it had
fallen back to the 15th ; in 1257 years, that is in 1582,
it fell on the 11th. Still the 21st of March, by the only
calendar in use at that time, was accounted the date of
the vernal equinox, by which date Easter was determined, so that, in 1582, when it was the 21st by the
calendar, the correct date was the 31st. Hence, the
error had been increasing at the rate of three days every
400 years until in 1582 it amounted to ten days.

Again it should be borne in mind that the Pope was
a churchman and wished to abide by the decision of
that council in celebrating the festival of Easter, so he
drops the ten days and restores the vernal equinox to
the 21st of March, its date at the meeting of the Council of Nice in 325, the date by which Easter day was
determined. He not only made the correction, but he
so reformed the calendar that the solar and the civil
year are now made to coincide very nearly. Had he
dropped the thirteen days, the vernal equinox would

have been restored to the 24th of March, its date in the time of Cæsar, and the 24th would still be its date. But the Council of Nice decided that the 21st should be the date by which Easter day should be determined. Hence the reason for dropping the ten days instead of the thirteen is evident ; and it is also evident that the Pope acted understandingly when he made the correction in 1582.

ERRATA.

On 51st page, ninth line from the bottom of the page, instead of $1453 \div 43 = 63+$, should be $1453 : 4 = 363+$.

On 76th page, twelfth line from the top of the page, 356 should be 365.

THE COLUMBUS CELEBRATION.

THE WELLSBORO MAN WHO BROUGHT ABOUT THE
CHANGE IN THE DATE.

State Superintendent D. J. Waller, in the School Journal.

Who brought about the change in the date? It was down in the books as October 12th. The Committee of the National Educational Association issued circulars to the country to observe that day. Congress solemnly resolved that that day should be celebrated throughout the land. The Commissioners of the World's Fair fixed upon October 12, as dedication-day, and sent out invitations to the exercises. Suddenly there was a change. Heralded by no newspaper discussion, preceded by no exhaustive treatise, without any authoritative decree, a change was made to October 21.

The following facts are indisputable. An aged retired minister in the Methodist Episcopal Church, living in Wellsboro, Tioga County, Pa., author of a little book entitled "Our Calendar," Rev. George Nichols Packer, saw the error. He possessed the confidence of Judge Henry W. Williams, of the Supreme Court of Pennsylvania. After laying the facts before him, he secured through him the approval by the several Justices composing that body, of an effort to change to the proper date, October 21st. Equipped with this approval, he secured the endorsement of his project by Governor Pattison and some of the heads of the Executive departments at Harrisburg. He then went to Washington, gained an audience with the President, laid the subject before the member of Congress from his district, and went before the Congressional Committee.

The evidence in support of the proposition was so presented that it could not be successfully disputed. Congressman W. A. Stone skillfully enlisted influential collegiates in an effort to correct the error already widely spread. The correction by the National Legislature was in time to have its influence upon President Harrison, who named October 21 in his proclamation, as the day to be observed, and Boston and Chicago fell into line.

All honor to Rev. George Nichols Packer, of Wellsboro, Tioga County, Pennsylvania.